D0425377

Mass Destruction

Illustrations

Acknowledgments

When I was a boy growing up in western Montana, my parents took me and my four brothers to see the Berkeley Pit in Butte, back when it was still a pit and not a toxic lake. I can still recall the mixture of awe, excitement, and unease I felt at seeing that shattered industrial landscape, one so different and seemingly distant from our home in the green and pleasant Missoula Valley. First thanks must go to my parents, then, for introducing me to some of the major fault lines that divide the modern world and sparking an enduring fascination with big pits.

Thank you to David Hounshell for his invaluable advice on an earlier version of this work. Many thanks as well for the assistance provided by the University of Delaware, the Hagley Museum and Library, the Smithsonian's National Museum of American History, the National Science Foundation, the Huntington Library, and the University of Wyoming's American Heritage Center. Many teachers, colleagues, and friends at all of these institutions helped out along the way, including Jeffrey Stine, Roger Horowitz, Reed Geiger, John Byrne, Philip Scranton, Jordan Kleiman, and Christine Skwiot, to name but a few.

Every writer should be so lucky as to find a home like the Department of History, Philosophy, and Religious Studies at Montana State University. In the department office, Diane Cattrell and Deidre Manry have been constant sources of good humor, wisdom, and assistance. Two generous Scholarship and Creativity Grants from the Montana State University Office of the Vice President for Research, Creativity, and Technology Transfer provided precious time for revisions and additional research and writing,

Acknowledgments

while the extraordinary climate of intellectual excitement within the department provided inspiration and renewal. The department's new doctoral program has attracted superb students whose insights during seminars and in their papers contributed to the book in many ways. Thanks to Michael Fox, Robert Gardner, Jerry Jessee, Robin Patten, Betsy Gaines Quammen, Paul Sivitz, Diane Smith, Bradley Snow, Constance Staudohar, and Wendy Zirngibl, as well as to the department's many excellent MA students who are too numerous to mention individually. My deepest appreciation goes to my department colleagues, who have provided so much intellectual and personal companionship: Jim Allard, Prasanta Bandyopadpadhyay, Gordon Brittan, Robert Campbell, David Cherry, Susan Cohen, Dan Flory, Kristen Intemann, David Large, Sanford Levy, Michelle Maskiell, Mary Murphy, Carla Nappi, Sara Pritchard (now of Cornell University), Michael Reidy, Robert Rydell, Lynda Sexson, Billy Smith, Brett Walker, and Yanna Yannakakis. Across the campus in the MSU Department of Physics, Shannon Willoughby provided some valuable feedback on my discussion of star formation and nucleosynthesis in chapter 2.

Thank you also to Doreen Valentine at Rutgers University Press for recognizing the potential of an early draft of this book and skillfully guiding it through to publication, as well as to the editing, production, and marketing staff at Rutgers for their fine work.

A small part of the material in chapter 3 appeared previously as "The Limits of Eco-Efficiency: Arsenic Pollution and the Cottrell Electrical Precipitator in U.S. Copper Mining," in *Environmental History* 5 (2000): 336–351, a publication of the Forest History Society and the American Society for Environmental History. The section in chapter 5 on tourism at open-pit mines was first published in a slightly different form by the Montana Historical Society as " 'See America the Bountiful': Butte, Anaconda, and the American Culture of Consumption," in *Montana: The Magazine of Western History* 56 (Winter 2006): 5–17. Thank you to the editors and publishers for their generous permission to republish this material.

Finally, I want to thank my family. Besides introducing me to Butte, my parents, Robert and Mary LeCain, have always been an inexhaustible source of support and encouragement. My brothers and their wives opened their homes to me too many times to count, keeping my western research travels both affordable and enjoyable. Special thanks to my wife, Cherí, daughter, Carina, and son, Daniel, who provide the foundation and the reason for all I do. In gratitude for their patience and support during the many hours spent apart while "Daddy is working on his book," it is lovingly dedicated to them.

Mass Destruction

ONE

In the Lands of
Mass Destruction

He, in his own age, was almost as important a figure
as Mr. Jobs is in our age.
—A friend of the Jackling House

The wooded hills and elegant homes of Woodside, California, at first
might seem an odd place to begin a book about mass destruction technology. On a warm early summer's evening in this fashionable community
south of San Francisco, the air smells of fresh-cut grass and eucalyptus. A
light breeze carries a hint of salt air from the ocean ten miles to the west
and stirs the leaves on the tall coastal oaks that blanket the hills a dense
green. Near the fork in the road where Robles Drive splits right from
Mountain Home Road, there is a large tree-covered estate with an abandoned Spanish Colonial mansion—or at least what was supposed to look
like an old Spanish Colonial mansion. It was actually built in the early
1920s for the copper mining tycoon Daniel Cowan Jackling, one of the
most important mining men of the twentieth century and a key inventor
of the modern technology of mass destruction. But standing near the
house in the fast-fading western light, you would search in vain for a historical plaque or any other hint that a man due some measure of remembrance lived much of his adult life here at 460 Mountain Home Road.
Dead now for more than half a century, Jackling seems to be as forgotten
and inaccessible as his deserted mansion, veiled and locked away amid the
fog and oaks of the Pacific coast.

1. Daniel Jackling built this elegant Spanish Colonial Revival mansion in the affluent community of Woodside south of San Francisco in 1926. Steven Jobs, the founder of Apple Computer, is the current owner. His efforts to tear down the mansion and replace it with a modern house have been stymied by historic preservationists. *Courtesy Woodside History Archives.*

For reasons that will become evident, it is difficult today to gain permission to visit the Jackling house. Yet old pictures show an impressive red-tile-roofed adobe house that sprawls over a large and elegant estate. Though the seventeen-thousand-square-foot home is not nearly so ostentatiously extravagant as William Randolph Hearst's San Simeon to the south, nonetheless its thick stucco walls enclose thirty-five rooms, thirteen baths, a custom-built Aeolian organ (a sort of player piano on steroids), and ubiquitous copper fixtures throughout—a nod to the original natural source of Jackling's wealth.[1] Inside, heavy dark oak beams support the ceilings, and decorative Spanish tiles grace the floors and walls. The effect is less authentic Spanish Colonial than 1920s Hollywood glitz. One suspects the fictional film star Norma Desmond would have found the home perfectly to her taste if it had been built on L.A.'s Sunset Boulevard.

Unfortunately, the years have not been any kinder to the Jackling house than they were to Norma Desmond. Unoccupied and largely neglected since the mid-1990s, today the mansion is showing signs of serious decay. The once brilliant white adobe walls are streaked with sooty black stains, and dry rot and mold have begun to invade the interior. In the music room the grand Aeolian organ still stands, its four tiers of black and white keys covered with layers of dust and dead leaves. Water pools on the Spanish-tiled floors, and the electric chandelier's faux candles bend at wild angles or hang broken from their sockets. Dank, gloomy, and slightly spooky, it would be a fine place to celebrate a memorable Halloween night. It is difficult now to imagine the famously elegant parties that Jackling and his wife, Virginia, once hosted here, to picture the movie stars, politicians, and other celebrities of the day talking and laughing beneath the brightly lit chandelier as the swelling music of the organ echoed through the tiled halls.

If the artifacts a society chooses to save from the past suggest what it wishes to remember, then those that it abandons to rot and decay may suggest what it would just as soon forget. Daniel Jackling died at his Woodside estate in 1956. Virginia followed a few years later. In the decades since, the house had several proud and doting owners and remained a much admired jewel of Woodside. By the early 1980s, though, the community found itself neighbor to the dynamic new economic hotspot of Silicon Valley, a boom that brought a new generation of the recently rich seeking convenient Arcadian sanctuaries. One of them was Steven Jobs, the computer whiz who made his fortune with Apple Computer. Jobs bought the Jackling estate in 1984 and lived there for more than a decade before he and his young family moved to a smaller home in nearby Palo Alto. Jobs eventually decided he wanted to move back to the spacious wooded estate, though not back to Jackling's quirky old manse. Instead, he proposed to tear down the Jackling mansion and replace it with a (relatively) smaller modern house. Jobs apparently stopped maintaining the house after about 2000, starting its decline into disrepair. Meanwhile, his plans to raze the mansion sparked controversy, and historic preservationists petitioned for an injunction to stop the computer magnate. In early 2006, a San Mateo County Superior Court judge agreed with the preservationists and ordered a halt to the project. Jobs, who now refers to the Jackling house as an "abomination," has offered to give it away to anyone who would move it to another site.[2]

The controversy over the Jackling house is hardly unique in a culture where the rush to embrace the future has long outpaced efforts to preserve the past. Yet the story also offers a deeper glimpse into the history of how Americans view nature, technology, and their own material environment, which is at the heart of this book. Even the preservationists fighting to save the house emphasize its architectural importance over its historical significance as Jackling's former home. George Washington Smith, a fashionable Santa Barbara architect famous for his Spanish Colonial Revival style, designed the house. An article in the local *Mercury News* captures Jackling's only secondary relevance: "It's protected because Smith was an important architect, even if most of his famous stuff is in Santa Barbara, and because Jackling was an important guy, even if most folks have never heard of him."[3] Even a letter from the California State Historical Resources Commission emphasizes the house's architectural importance, while adding almost as if an afterthought, "The property also derives its significance from Smith's clients, Daniel and Virginia Jackling."[4] Indeed, one journalist concludes that Jackling's connection to the house would best be forgotten altogether: "It's time to start calling it the Jobs House instead of the Jackling House."[5]

In the early years of the twenty-first century, it seems, the man whose technology provided the nation with billions of pounds of cheap copper was less worthy of remembrance than an upscale Santa Barbara architect who specialized in faux Spanish Colonial buildings. While Jobs and the preservationists may have fought over the architectural significance of the house, they seemed to agree that Jackling and his accomplishments were largely insignificant. In such careless or willful acts of historical amnesia are suggested the destructive ironies of modern technological America and its bizarre and often dysfunctional relationship to nature. Only a society utterly secure in its comfortable technological environment could so easily ignore its material foundations in the natural world. Earlier generations of Americans, living at a time when modern dichotomous views that separate humans and technology from a pristine concept of nature as wilderness were still emerging, had actually been much quicker to recognize the connections. During the first half of the twentieth century, American presses published numerous books on the history, extraction, and significance of critical modern raw materials like copper. Though sales and circulation figures are difficult to determine, presumably these publishers believed that a sizeable number of Americans were interested in these topics. The books were all uncritical in their celebration of the accomplishments of modern mining, and few mentioned the tremendous costs in both human and environmental losses. Still, in comparison to the modern distancing between the built material environment and the natural world, these earlier works are refreshing in their clear and unapologetic explanations of how technological society had its roots deeply planted in nature.[6]

In the Lands of Mass Destruction

Nearly a century ago, Thomas Edison was presented with a solid cubic foot of copper as a symbol of the natural raw material that had made his system of electricity possible—a story we will return to later in this book. In contrast, consider the irony of Steven Jobs today, a man whose fortune is also built on electronic machines—machines filled with copper parts and powered by electricity distributed on copper wires—attempting to tear down the house of a man whose own fortune came from providing that very copper. Perhaps Jobs is aware of this contradiction. If so, it does not seem to have influenced his feelings about the house. Regardless, this story of a California real estate battle points toward the way modern technological society often keeps us from recognizing our everyday dependence on raw materials extracted from nature.

As one of the relatively rare champions for preserving the mansion for its historic significance rightly notes, Jackling, "in his own age, was almost as important a figure as Mr. Jobs is in our age."[7] Perhaps this forgetting of connections and history is one of the ways we make the sacrifices demanded by modern industrial civilization a bit less painful. For Steven Jobs, minimizing Jackling's contributions to the modern world may be tactically useful, even if it means ignoring the fundamental connection between his own industry and Jackling's. But Jobs is hardly alone. It has long been psychologically useful for all Americans to ignore our own connections to the technology of mass destruction that provided us with cheap and abundant copper—the copper that carries the electricity that powers the computer used to write these words. The ease with which Jobs or any of us may forget or ignore our fundamental connections to the natural first source of our material world is a problem that lies at the heart of this book's story.

TWIN SONS OF DIFFERENT MOTHERS

To be fair, both Jobs and the preservationists were not unusual in their historical amnesia: Daniel Cowan Jackling may well be one of the more important men of the twentieth century that no one has ever heard of. Well before Jackling died in 1956 at his Woodside mansion, he had established one of the world's greatest copper mining companies, accumulated a sizeable personal fortune, and received numerous awards and honors. Decades earlier, President Woodrow Wilson had personally awarded Jackling the Distinguished Service Medal for his national contributions during World War I. In 1933, he received the John Fritz Medal, one of the engineering profession's highest honors and one Jackling shared with the more famous 1929 recipient, Herbert Hoover, the only mining engineer ever to become president. Nonetheless, Jackling's death passed largely unnoticed by most busy midcentury Americans. The *New York Times* ran a glowing though relatively brief obituary, and even his professional colleagues failed to really recognize and commemorate the true significance of Jackling's work. The relative silence was partly just a consequence of longevity: at the age of eighty-six, Jackling had simply outlived what fame he had once enjoyed.[8]

Americans are supposedly notorious for their historical ignorance, yet time and forgetting cannot fully account for Jackling's decline into obscurity. Contrast Jackling's passing with that of his close contemporary Henry Ford, a man whose own fame and fortune were founded on related (though not identical) principles and depended at least in some small part on Jackling's work. Born six years before Jackling, Ford lived nearly as long, and the two men passed through almost exactly the same stream of history. When Ford died in 1947 at the age of eighty-three, Americans and much of the rest of the world commemorated the Detroit automobile maker as if he were a beloved statesman or an honored war hero. Harry Truman, Winston Churchill, and even Joseph Stalin sent word of their deep admiration for Ford. Newspapers ran above-the-fold front-page stories.[9] The *New York Times* printed a long and adulatory obituary. Ford's career, the *Times* noted, "was one of the most astonishing in industrial history" because of "the revolutionary importance of his contribution to modern productive processes."[10]

Admittedly, at first glance a comparison between Ford and almost any other industrialist (except perhaps Thomas Edison) may seem inapt. Ford was one of the most famous businessmen of the twentieth century. Few could ever match the worldwide notoriety of the creator of the Model T, the assembly line, and modern mass production. Millions shared an oddly personal connection with Ford, thanks to their mass consumption of his company's cheap and reliable automobiles. He was the public symbol of a machine and a lifestyle many adored. However, it is precisely these two accomplishments—mass production and mass consumption—that should have closely linked Ford and Jackling in the public imagination. Jackling's accomplishment was to pioneer a third and closely related system, without which the other two would have been all but impossible: mass destruction, a technological system for cheaply extracting huge amounts of essential industrial minerals from the earth's crust.

Still, Jackling did not manufacture mass quantities of a popular consumer product like a car. Nor did he help to create the cultural meaning of such products through industrial design, advertising, and the other tools of modern mass consumption. Rather, Jackling's achievements were made in a much less public and prominent field, in an endeavor that has for centuries remained distant and hidden from the more obvious world of the urban factory or the boisterous marketplace. Jackling was a copper miner—a provider of one of the most important natural elements of the modern age. He perfected a technology capable of supplying immense quantities of copper that manufacturers subsequently transformed into wires for countless miles of power lines and parts for billions of cars, radios, and refrigerators. By midcentury, more than 60 percent of the world's entire production of copper was extracted using Jackling's system of mass destruction.[11] Other engineers adapted the technology to different minerals, and the same basic principles emerged in extractive industries as varied as logging and fishing.

I have chosen the term "mass destruction" to describe Jackling's technology only after much thought and debate, and use it advisedly. For most people, of course, these words evoke the mass destruction of human life and property, either through war or terrorism, and they are usually preceded by "weapons of." This is not precisely the meaning I wish to convey here, though I will argue that the past century's technological and ideological "advances" in the efficient destruction of humans were not unrelated to the efficient destruction of mountains. Indeed, this murderous sense of the phrase, with its echoes of Guernica, Hiroshima, and cold war Armageddon, may help to explain why it has not previously been used to define this type of mining. What mining engineer or company would want to suggest that their extraction technology was in any way analogous to weapons of mass human death?

No other phrase, however, better captures the essential traits of this transformative but often overlooked technology that was a necessary condition to the building of the modern industrial and postindustrial world. The term is also appropriate because it echoes the better-known concepts of mass production and mass consumption—both of which depended on mass destruction to supply the essential raw materials. As the antonym to the phrase "mass production," "mass destruction" suggests this close association, although from a strictly physical standpoint, mass production no more "produces" new materials than mass destruction "destroys" them. Both processes merely rearrange matter into new forms. Some might suggest the phrase "mass extraction" as a less pejorative label for the technology, but that term fails to convey one of the key properties of modern open-pit mining: its sheer destructiveness. Large-scale mass extraction of ores can be achieved without mass destruction. Big underground mines where skilled miners carefully select only the richest ores can produce massive amounts of minerals from some types of ore deposits. These mines merit the term "mass extraction."

Mass destruction was not, however, just a matter of size. Rather, as perfected by Jackling and others it was a means of increasing the *speed* of mining and thus achieving a certain narrowly defined type of efficiency. The enormous size of the operations was a necessary adjunct to using coal- and oil-powered machinery to dramatically increase the rate at which ore was extracted, moved, and processed.[12] As a result, mass destruction mines were so efficient that they could profitably mine ore deposits that would otherwise have been worthless. The phrase "mass extraction" also suggests a misleadingly neat, precise method of carefully removing ores, rather like a dentist extracting a rotten tooth. "Mass destruction" better captures the sheer explosive messiness of big open-pit mining and thus conveys some of the inherent environmental destructiveness of the technology. While I do not wish to exaggerate this destructiveness, neither do I wish to perpetuate the cultural tendency to ignore it—an example of yet another of those inconvenient truths that lay at the heart of modern material affluence and its accompanying "democracy of goods."

Daniel Jackling did not single-handedly invent mass destruction mining any more than Henry Ford single-handedly invented mass production. Ford, though, became the symbol of a technology and an era, while Jackling has been all but forgotten. Would Steven Jobs have even contemplated tearing down the Henry Ford house? Modern Americans, and perhaps people around the world, have generally been far less interested in remembering the providers of basic raw materials than the producers of shiny new objects of desire. This was not always the case. An earlier era once celebrated mining men as the keys to modern progress, and appropriately so. The story of the Jackling house, though, reminds us that one of the defining features of modern technological society is its tendency to distance us from the environmental first sources of our material world. As the historian William Cronon argues, the late nineteenth-century development of centralized corporate meatpacking in Chicago made it much easier for consumers to forget the living natural animals who became their dinner. "Meat," Cronon writes, "was a neatly wrapped package one bought at the market. Nature did not have much to do with it."[13] In mining, the connections between natural source and consumer product are arguably even more obscured and the moral questions even less evident.

In the Lands of Mass Destruction

Americans have a prejudice for production, finding tales of clever mechanical invention, production, and consumption more interesting than
those chronicling the sources of raw materials. To a lesser degree, the prejudice has affected historians as well, and thus the stories we tell about the
past. Consider one popular recent college survey textbook, *Inventing America: A History of the United States*, an otherwise superb text that emphasizes
the importance of technological developments in the nation's history. Despite its focus on technology and industrialization, the text makes only
passing references to the role of mining in post–Civil War American history. As in most American history surveys, the history of mining technology and business is relegated to a few sentences in a section surveying the
late nineteenth-century development of the American West.

By contrast, *Inventing America* devotes nearly two pages to the story of
Henry Ford and the automobile and includes pictures of Ford with his first
automobile and of an early assembly line. Students reading *Inventing America* will learn that Ford's success lay in "improving the techniques of mass
production" and that by 1920 factory sales of automobiles had risen to the
mass consumption levels of 1.9 million a year. In the same section, they are
also told how "homes and workplaces were lit by electricity," and men and
women "communicated with one another through the telegraph or over the
phone." Automobiles, electrical power grids, telecommunications—all the
magnificently modern technological systems of mass production and consumption are powerfully and appropriately evoked. The words "copper"
and "mining," however, make not even the briefest of cameo appearances,
and they will scarcely appear again in the subsequent four hundred pages.

Copper and mass destruction mining did not cause the tectonic societal shifts of the modern era nearly so clearly as electric lights, cars, and the
other technologies that used copper. The connections backward from a
stylish new refrigerator with copper coils to a mine in Utah are not at all
obvious. It is precisely this disconnect between human products and the
environmental source of raw materials—between what we label "technology" and what we label "nature"—that needs to be closed if we are to better comprehend the dynamics of the modern world. All technological devices, from steam engines to computers, are made from nature and use
natural properties and principles to operate. As Albert Einstein explained
a century ago, humans neither create nor destroy matter but merely transform it from one state to another. In this sense, "nature" is always around
us, and the separation between technological and natural environments
can be seen as a powerful but misleading illusion.[14] The astounding ability

of modern technology to reshape nature in so many ways creates the illusion that our world of cars, houses, and skyscrapers is utterly of our own devising, removed and separated from nature. As a result, many Americans have even gone to the opposite extreme and come to believe that true nature exists only in wilderness areas supposedly devoid of human effects—a view that Cronon has famously suggested was "getting back to the wrong nature."[15]

One goal of this book is to examine ways in which we might instead find our way back to something more like the "right nature," to a concept of nature that more clearly includes humans and their technologies in its definition. As scientists have shown in recent decades, there are really no environments left on the planet completely free of anthropogenic effects. Cores drilled in the Greenland Arctic ice sheet show traces of widespread atmospheric lead contamination from early Greek and Roman ore smelting, peaking between 500 BCE and 300 CE.[16] Two millennia later, greenhouse gases generated by human activities now affect nearly every ecosystem on the planet. Human technological systems have become so thoroughly integrated into environmental systems that there is nowhere on earth that has completely escaped our influences. It no longer makes sense to draw clear lines between technological and ecological systems. This is not in the least, of course, an argument against identifying and preserving areas where the human mark on the land remains less evident. The differences between Montana's Bob Marshall Wilderness and the city of Los Angeles are obvious. But for all its value, wilderness preservation offers no real solutions to the modern dilemma of managing a world of more than six billion people, most of whom would likely choose cheap and readily available electricity, cars, and refrigerators over wilderness preserves.

Fortunately, the choice need not be a zero-sum game, but only if we begin to see much more clearly how our technological environment is inextricably linked to our natural environment. Indeed, we would do better to learn how to think of the two as a unified whole or an "envirotechnical system," both because this reflects physical realties and because this analytical approach offers greater insights into preserving the best aspects of both wilderness and civilization. Consider how differently we might think if we could learn to recognize that nature was always around us, if every schoolchild knew the basic natural history of the metals in a car or the wood fiber boards in a house. Might the day ever come when we could see in a rap star's thick gold necklace not just a flamboyant cultural symbol of wealth and power but also the 360 tons of rock mined to produce the gold?

Or might an outdoor science school offer, in addition to discussions of the hunting habits of the great horned owl and the role of fire in serotinous pine tree propagation, lessons on the complex natural and social history of the trucks and SUVs that likely brought many of the students and teachers to the camp? In the modern envirotechnical world, developing a far deeper understanding of the technological may be the only realistic path to saving the natural.

BIG MINES, TALL STACKS

Though increased speed is at its heart, mass destruction mining is also big—so big as to defy easy description. Commentators often grope for words to convey the scale of open-pit mining operations, though neither words nor photos are adequate to the task. A favorite device to capture the stunning size of Daniel Jackling's Bingham Pit copper mine near Salt Lake City, Utah, is to note that it is one of only two man-made objects that can be seen by astronauts from outer space, the other being the nearly four-thousand-mile-long Great Wall of China. Dozens of articles, Web sites, and tourist guides reiterate this space-age superlative, even though a bit of thought suggests it is obviously not true. If the fifteen-foot-wide Great Wall is visible to space shuttle astronauts, who orbit at about 135 miles above the earth, then so are countless highways, airports, and dams. Likewise, scores of other pits and major excavations (think of the Panama Canal) around the world are equally visible at that altitude. Even the Bingham Pit is no longer the largest open-pit copper mine in the world, as that dubious honor has recently passed to the Chuquicamata Pit in Chile. Still, minor quibbles aside, the Bingham Pit unquestionably remains among the biggest single human-made artifacts on the planet. For a crew of astronauts on a voyage to Mars, the pit might well be among the last human-made features they could make out as the planet slowly receded in the distance. A fitting final symbol, given that the astronauts' technological home away from home in space would likely contain copper, aluminum, gold, and other metals mined in open pits.

Accurate or not, the popularity of using the Great Wall of China as a yardstick suggests the difficulty humans have in coming to terms with the size of the Bingham Pit and other products of mass destruction mining. Confronted with such a huge artifact—one that is literally beyond the normal human sense of scale and proportion—people simply grasp for analogies to the biggest thing they have ever heard of. Or is it that there is

2. Jackling's Bingham Pit mine south of Salt Lake City, Utah, in 1972. Currently three-quarters of a mile deep and two and a half miles wide, the Bingham Pit is one of the largest human-made artifacts on the planet. Each of the steplike benches is between fifty and eighty feet high. The tiny black dot slightly above the electric power pole on the left is an immense twenty-foot-tall steam shovel. *Library of Congress, Prints and Photographs Division, Historic American Engineering Record, Reproduction Number HAER UTAH, 18-BINCA, 1–1.*

something comforting in the comparison to a very long but nonetheless distinctly human-sized wall constructed centuries earlier, suggesting that the pit may not be such a jarring departure from past human experience after all? If so, the sense of comfort is unwarranted, as the Bingham Pit and its leviathan cousins far exceed the scale of any other single discrete human-made creations from the pre-twentieth-century era.

The best way to comprehend the Bingham Pit is to go there. Kennecott Utah Copper, a subsidiary of the international mining company Rio Tinto, owns and operates the pit today, and the company encourages visitors. The Bingham Pit viewing stand and visitors' center have been popular tourist attractions for decades. Looking southwest from central Salt Lake City, a visitor can easily spot the pit as a flat reddish-yellow gouge into the eastern flanks of the Oquirrh Mountains that run due south from the city and lake. It looks as if some massive explosion had annihilated several of the

Oquirrh peaks, leaving the range with a yellow-stained gap where the surrounding topography insists a forested mountain should be. Kennecott charges tourists a modest five dollars (all proceeds donated to charity, school buses are free) to drive up the snaking canyon road to the pit overlook. At 6,700 feet, the overlook is chilly even on a sunny day in early May, and gusts of gritty wind snap at jacket sleeves and the big American flag flying above the visitors' center. Undeterred, a few dozen people lean against the concrete barriers to peer down into the depths of the big hole, looking just like tourists gaping in awe at the Grand Canyon—though it is worth recalling that the Colorado River required millennia to carve its canyon, while the Bingham Pit is barely a century old.

Numbers may be inadequate to capture the scale of the Bingham mine, but they are nonetheless impressive. At its widest point, the pit stretches more than two and a half miles from rim to rim. From the highest point to the bottom, the pit is three-quarters of a mile deep, and the average depth is half a mile. Over the past century, more than five billion tons of rock and ore have been blasted out and removed from the mine. If the Burj Dubai skyscraper in the United Arab Emirates, the tallest building in the world as of 2008, were to be lowered into the pit, it would not even reach halfway up to the rim. Should the world ever decide to hide its major skyscrapers, the Bingham Pit is well up to the task.

If visitors to the pit take the time to tour the company museum, they will learn that the Bingham Pit was the creation of Jackling and his Utah Copper Company. A short film briefly explains the technological steps by which the low-grade ore in the pit is transformed into the pure copper for electrical wires, brass, bronze, and many consumer and industrial products. In a final dramatic flourish, curtains hiding a wall of windows open so the audience can once again look out over the pit that has provided so much of the material wealth of the nation. Neither the exhibit nor the film dwells on the environmental costs of the Bingham Pit, instead stressing Kennecott's attention to the issue in recent years. Massive amounts of copper, the exhibits seem to suggest, can now be had at little or no serious environmental cost. It perhaps goes without saying that Kennecott does not use the term "mass destruction" to describe the technology for making such impressive holes in the ground.

Jackling perfected the technology of mass destruction at Bingham partly because it was an interesting engineering challenge and partly because he wanted to achieve success and wealth—seventeen-thousand-square-foot mansions do not come cheap. But as free-market zealots are fond of pointing out, sometimes the self-interested pursuit of wealth actually can serve a greater public good, and the Bingham Pit was also an immense technological fix for what some saw as an impending crisis. During the early twentieth century, many mineral analysts had raised frightening alarms, predicting that the nation was on the verge of a severe copper shortage. Just as the process of national electrification was shifting into high gear, the well-known high-grade copper deposits in Montana, Arizona, and Michigan were beginning to decline. Jackling thought he had a solution to the crisis at Bingham: not the discovery of a new deposit—geologists had long known that there was an immense deposit of copper at Bingham— but the creation of a system for mining Bingham's low-grade deposit of disseminated ore. Mining experts had previously dismissed the Bingham deposit as worthless; all the numbers showed the copper would sell for less than it cost to mine. Jackling agreed this was true, *if* the big deposit was mined using conventional underground mining technology. But what if the speed of mining was greatly increased by digging down from the surface using high explosives and steam shovels to create a giant open pit?

This was Jackling's legacy to the world: a system of mass destruction mining as powerful as the system of mass production that Henry Ford and others were developing at almost precisely the same time. Jackling's idea could only have worked in the era of powerful but crude engines, energy-hungry steam, electric, or diesel monsters that could do the work of thousands of men or animals in a fraction of the time. Hydrocarbons, and later cheap hydropower, were the food that kept the mechanical muscles moving. Jackling's system demanded innovation at every stage of mining in order to speed the movement of raw copper ore from pit to concentrator to smelter. When it was all in place, the mountain of previously worthless rock at Bingham had been transformed into a bonanza copper mine that would eventually produce billions of dollars of wealth. The mine became so valuable that Jackling could build his Woodside estate, where he might sit in the elegant gardens and listen to his mighty Aeolian organ, perhaps congratulating himself on having helped to avert an impending copper crisis.

Although Jackling considered himself a conservationist of the efficient-use school, there is little evidence he gave any thought to the environmental costs of his achievements, which extended far beyond the obvious damage caused by the big pit itself. Indeed, the full consequences of open-pit mines cannot be understood without also studying the closely linked ore processing and smelting technologies. Mass destruction was a system of tightly integrated technologies, and the environmental effects of getting the copper out of the ore were often even greater than those of getting the ore out of the mine. Jackling pioneered the open pit, the real core of mass destruction, but he borrowed much of the subsequent ore processing technology from other copper mining operations. To understand these innovations and their consequences will require visits to other regions in the landscape of mass destruction as well. Put simply, big copper mines demanded smelters with tall smokestacks, and while Bingham had the biggest pit, Montana had the tallest stack.

To the casual traveler speeding by at eighty miles an hour, the Deer Lodge Valley must at first present the picture-perfect view of an idyllic rural Montana landscape. Cattle graze on rolling pastureland, green irrigated farms abound, and stately old cottonwoods crowd the banks of the winding Clark Fork River. Only the unusually sharp-eyed might notice the occasional patches of barren land and oddly colored soils scattered among the fields and river bottoms, a few brief dark notes in an otherwise cheery pastoral symphony. But as the road moves on past the small community of Deer Lodge, even the most inattentive will at some point spot the looming dark tower of the Washoe smelter smokestack in the distance, rising at the far southern end of the valley and looking rather like a photo-negative image of the Washington Monument. Move closer and the astonishing size of the 585-foot-tall masonry stack (the tallest of its kind in the world, and thirty feet taller than the Washington Monument) becomes apparent. As the highway skirts to the east of the stack, dusty barren hills border the road for nearly four miles. These are the forty-foot-high remnants of the giant Opportunity tailings ponds, industrial middens created by a century of copper processing.

A few miles later the highway turns east, the mountains crowd in, and the incongruous postindustrial landscape of the Deer Lodge Valley disappears in the rearview mirror. But drive another fifteen minutes up into the higher and smaller Summit Valley and the raison d'être of the Washoe smelter stack appears, an immense scar carved out of the town of Butte: the Berkeley Pit. Before Berkeley, the Anaconda Copper Mining Company obtained its ore from deposits beneath the city of Butte tapped by seven main shafts and thousands of miles of tunnels, some built nearly a mile beneath the surface. Beginning in 1955, however, Anaconda followed the example of Jackling's Bingham mine and shifted most of its operations to the mass destruction surface operation that became the Berkeley Pit. By the early 1980s, the twisting oval hole was almost 1.5 miles wide and 1,800 feet deep and had consumed a sizeable chunk of the city.

In 1982, the Atlantic Richfield Company (ARCO, Anaconda's corporate successor) shut down all mining in Butte and turned off the giant pumps more than half a mile below the city that had pumped ground water from the mines for decades. Soon the first puddles of water formed in the pit bottom. Then the puddles became a pond. As the groundwater continued to rise, resuming its previous natural level, citizens of Butte with a penchant for dark humor rechristened the flooding pit "Berkeley Lake." Some even staged a mass Hawaiian hula dance on its rim. But forget any images of cool limestone quarry swimming holes, much less white-sand Hawaiian beaches. The Berkeley Pit is a giant drain hole, the resting place of groundwater steeped through thousands of mile of subterranean passages, creating an acidic cocktail of heavy metal poisons. When a flock of migrating snow geese landed on the "lake" in 1995, more than 340 made the fatal mis-

3. The Berkeley Pit in Butte, Montana, began to flood after the underground water pumps were turned off in 1982. Currently over a thousand feet deep, the pit water is nearly as acidic as battery acid and contains a toxic brew of heavy metals. It must be constantly pumped and treated to prevent the further contamination of surface and groundwater reservoirs. *Photo by Timothy J. LeCain.*

take of lingering in the pit water for several days. Reportedly, the recovery of their carcasses was slowed because the birds' brilliant white feathers had stained the reddish-orange of the acid-laden water.[17]

As of 2007, the Berkeley Pit lake was more than nine hundred feet deep, and it is only the most obvious feature of the nation's largest Superfund site. The site also includes the immense underground mine workings, the nearby Anaconda smelter site and Opportunity tailings ponds, and a 126-mile stretch of the Clark Fork River. Cleanup efforts are under way and much progress has been made, but the task is daunting. At times, the only hope for a better future for Butte seems to come from "Our Lady of the Rockies," the brilliant white ninety-foot-tall steel statue of the Virgin Mary that benevolently gazes down over the town and pit from the high rocky spine of the Continental Divide.

The city of Butte has long inspired both awed contemplation and angry condemnation, whether because of the scale of the mining operations, the attendant environmental destruction, or some confused mixture of the two. Many of the historic political, economic, and labor upheavals that have periodically shaken this high northern Rocky Mountain city have been equally outsized, and almost all of them related in one way or another to the mining. In recent years, though, visitors are most likely to see the scarred landscape of Butte as evidence of some sort of environmental "original sin," the product of a greedy and rapacious mining company that stripped Butte of its mineral wealth and left behind a hollowed-out husk of a city that threatens to dry up and blow away in the high mountain winds. The author of one article on the environmental problems caused by the copper mining suggests the evocative imagery of "Pennies from Hell."[18] Another refers to the Anaconda as "the Snake" that killed the snow geese.[19] Another, perhaps a bit desperate to find an adequately nefarious analogy, concludes that the old Washoe smelter stack reminds him of the tower of Isengard, the fortress of an evil wizard in J.R.R. Tolkien's *Lord of the Rings*.[20]

Whether as "the richest hill on earth" or as home to the nation's biggest Superfund site, Butte has always inspired hyperbole. Condemning the Anaconda has also long been a popular Montana pastime, and the condemnation is generally well deserved.[21] A ruthless (perhaps even murderous) opponent of the mine workers' long struggle to unionize, an unapologetic manipulator of Montana politics, and a censorious master of much of the state's media, Anaconda ran Montana like a corporate fiefdom for a good part of the twentieth century. Given the Anaconda's crimes against Montanans, it has been tempting to explain the environmental carnage at Butte and Anaconda as simply yet another example of "the Snake's" irresponsible behavior. That story makes for good drama, and its lessons about the dangers of corporate hegemony and capitalist exploitation and commodification of nature are important and generally well understood. However, new evidence and theoretical approaches have increasingly suggested we should resist a too facile framing of Butte's environmental history as merely a morality play on the evils of early twentieth-century corporate capitalism. While containing elements of truth, such a simple declensionist tale tends to obscure other important aspects of the story that offer more useful lessons for understanding the human relationship to the environment. This other story that needs telling is one of arrogant overconfidence more than deliberate malice, of difficult trade-offs more than moral absolutes, about shared guilt rather than convenient scapegoats. Above all, it is a story about the dangers that come from thinking that humans and their technological systems are separate from nature and its ecological systems, and about the power and wealth that came from and propagated such illusions.

Jackling's pit would be the ultimate expression of this modern conceit, but to understand its roots demands probing deeper into the past and into the underground. Precisely because it initially demanded the creation of an underground world that was seemingly unnatural and separate from the living world above, the transition from underground to open-pit mining offers the perfect case study of this illusion that humans and their technological systems are distinct and separate from nature.

DEPTH ANALYSIS

In 1932, the cultural critic and historian Lewis Mumford went "underground" in the elaborate life-size mining exhibit at Munich's Deutsches Museum. He was so moved by his virtual experience of underground space that he later made mining and minerals a central theme in his influential 1934 history of the machine age, *Technics and Civilization*.[22] His basic argument was straightforward: mines and mining were instrumental to the process of early capitalist industrialization through their provision of coal, iron, and the other raw materials for power machinery. Mumford went beyond this, though, to argue that the mine was also a perfect allegory for what he viewed as the increasing artificiality of technological civilization, its distancing from the organic rhythms of the natural world. "The mine," Mumford writes, "is the first completely inorganic environment to be created and lived in by man." Fields, forests, and oceans, Mumford argues, "are the environment of life." By contrast, "the mine is the environment alone of ores, minerals, and metals. Within the subterranean rock, there is no life, not even bacteria or protozoa, except in so far as they may filter through with the ground water or be introduced by man."[23] Elsewhere in the book, Mumford elaborates the theme, noting that in mines, "day has been abolished and the rhythm of nature broken: continuous day-and-night production first came into existence here. The miner must work by artificial light even though the sun be shining outside; still further down in the seams, he must work by artificial ventilation, too; a triumph of the 'manufactured environment.'"

Mumford suggests a compelling if misleading idea here: the underground mine as a concrete expression of capitalist environmental and social relations. The unnatural inorganic environment of the mine, he concludes, fostered the equally unnatural and exploitative regimentation of mine workers. Capitalism and its destructive technology have broken the "rhythm of nature" every bit as much as a dead underground mine. There is much of value in Mumford's analysis, and this book will return to some of these same themes of artificial light and ventilation in a "manufactured environment." Likewise, Mumford is at his brilliant best in recognizing the importance of the mine as a metaphor for the modern world. As Rosalind Williams shows in her book *Notes on the Underground*, there are few spatial concepts seemingly more fundamental in human cultures than those that distinguish between "above" and "below," "inside" and "outside." Indeed, Williams argues that the fundamental spatial nature of the underground mine is the source of its appeal as a metaphor for modern existence: "It is the combination of enclosure and verticality—a combination not found either in cities or in spaceships—that gives the image of an underworld its unique power as a model of a technological environment."[24]

The unique physical nature of the subterrestrial mine, its sense of being an enclosed vertical world utterly apart from the normal human terrestrial worlds, explains its appeal to Mumford and others. In this view, the mine serves as the extreme expression of all that is human-made and therefore believed to be unnatural: the city, the factory, and the railroads and machines that invade a pristine countryside. Just as all human beings easily recognize the difference between the surface and the underground, so too is the difference between the natural and the technological supposed to be equally obvious and perhaps even instinctive.

Unfortunately, as is often the case with *Homo sapiens*, what is obvious and instinctive is often wrong, and that is the case here as well. Indeed,

Mumford's argument was precisely backward. It was not the artificiality of the mine that led to environmental devastation, but rather the seemingly commonsense belief that Mumford himself was guilty of further propagating: that human beings and their technologies could ever be separated from the natural world.

One of the most destructive and dangerous ideas of the past century was that Americans (and others) could engineer a technological world largely independent from the natural world, whether that be a mine, a city, or a controlled and isolated industrial dead zone. Nature, of course, would still be the source of raw materials, agricultural products, water, and air, as well as a dump for the waste products of industrial civilization. Increasingly, however, these natural systems were believed to be mere cogs in the larger technological system, distinctly secondary subsystems that could be fully controlled, rationalized, and engineered for maximum productiveness. When problems or unanticipated consequences arose in the technological incorporation of these natural subsystems, such as declines in productivity or harmful pollutants, engineers and scientists offered new approaches and technological fixes that promised to either repair the system or erect an effective barrier between human society and any adverse natural effects. Either way, the technological appeared so powerful as to be nearly independent from the natural. Not coincidentally, the opposite idea that real nature occurred only in wilderness areas supposedly untouched by human technology emerged at roughly the same time, further deepening the illusory divide between the human and nonhuman world.

Having supposedly extricated themselves from nature, modern humans could think of the natural world they had left behind as a resource to be bent to their will, to be simplified, rationalized, and optimized to maximize human wealth, power, and safety. The many and undeniable accomplishments of modern science and technology were the result. Yet, as the political geographer James Scott demonstrates, this "high modern" technological regime also contained the seeds of its own failures. Both the natural and technological systems they had so confidently engineered proved far more complex than imagined, prone to unanticipated reactions and consequences. Simplified monocrop forests collapsed. Smog choked the life from cities. DDT killed songbirds and turned up in mother's milk. At the same time, the human-natural divide was used to justify and "naturalize" a host of other injustices, from the dominance of the supposedly "rational" male over the more "instinctual" female to the rule of the technologically advanced Western nations over colonial peoples viewed as closer to nature and thus "backward." As the environmental historian Richard White rightly notes, "who gets to define nature is an issue of power with consequences for the lives of working people, Indian people, and residents of areas defined as wild."[25]

Given the many potentially harmful effects of the dichotomous view separating the natural and artificial, the ecological and technological, any effective analysis of environmental and technological history must take pains to avoid validating this modernist illusion. Precisely because mining has so often been seen as inherently unnatural, it actually offers an ideal opportunity for exploring a different way of thinking about humans and nature. Rather than viewing the subterrestrial mine as the antithesis of the terrestrial "environment of life," as Mumford's dichotomy suggests, it is far more useful to think of the mine as a simplified environment. While less organic than many (though perhaps not all) terrestrial environments, it is certainly no less natural than the aboveground world, and modern science now even recognizes the existence of a subterrestrial biosphere.[26] Nor is the subterrestrial work of miners somehow less natural than the work of the farmer. Rather, in this concept of the mine, nature still very obviously exists, but it exists as part of a hybrid ecological and technological mining system. Significantly, this mining system is somewhat less complex and more easily managed than that of an industrial farm or a tree plantation. The mine is thus a striking example of a constructed human environment created by experts (engineers, managers, skilled miners) through their knowledge and limited mastery of subterrestrial space and ecology. Having developed the tools to measure, map, and control the underground world, these early "environmental engineers" created deep subterrestrial spaces that were seemingly distinct from traditional terrestrial human environments—the first such environments in human history. They were in part "manufactured environments," but Mumford erred in suggesting they were the antithesis of the organic, natural, living environments of the terrestrial world. To the contrary, they were inseparably linked to these terrestrial worlds, just as a city is linked to countryside.

In recent years, scholars studying the interactions between technology and the environment have begun to develop other analytical methods that avoid lapsing into these convenient but false dichotomies.[27] In January of 2000, several historians who had intellectual and professional affiliations with both the Society for the History of Technology and the American Society for Environmental History established Envirotech, a special-interest group for scholars working in both disciplines. Many of the scholars working in this and related areas have begun to develop a unique new method of envirotechnical analysis, an approach that challenges the misleading distinctions between humans and their environment. Indeed, as the historian Edmund P. Russell argued during a recent conference roundtable on the subject, "all technology is made out of nature," while nature constantly feeds back into human technological and social systems.[28]

Such an envirotechnical analysis does not deny the value or necessity of sometimes using more conventional categorical distinctions between technology and environment, human and nature. Indeed, our language itself is constantly forcing us back into these dichotomies. Even to use the term "nature" is immediately to raise in the minds of most (at least Western) listeners a concept defined by whatever is not human, not technological.[29] English offers no good word for that which is distinct from humans and thus affected by their technologies, but which also simultaneously includes humans and recognizes their technologies as being both derived from and embedded in nature. Absent such a much-to-be-desired word, to argue that "all technology is made out of nature" or that "humans are a part of nature" inevitably tends to re-create the old dichotomies even while the intention is to collapse them. The goal of envirotechnical analysis, then, is to demonstrate how this system that is both human and nonhuman, artificial and natural, technological and ecological, does actually exist even if our culturally constructed ideas and words often keep us from recognizing it.

The technologies of mass destruction that created the Bingham and Butte mines, their gigantic smelters, and the nation's largest Superfund site are ideally suited to just such an envirotechnical analysis. Here misleading dichotomies abound: surface and mine, terrestrial and subterrestrial, organic and inorganic, natural and artificial, environmental and technological—even female and male. Look closely, though, and these seemingly solid categories begin to melt into air, not only in a metaphorical or semantic sense, but also in a strikingly concrete and physical way. From an envirotechnical perspective, Bingham, Berkeley, and other such pits of mass destruction emerge from the depths of cultural and historical misunderstanding to reveal their true "nature" as enduring physical manifestations of the tremendous powers and the tragic limitations of the modern ideologies, societies, and economies that created them. Put simply, the pits of mass destruction are the embodiment of a human cultural, economic, and technological relationship with nature gone badly awry. Worse, as the size of these pits continued to expand and it became possible to profitably exploit even minuscule percentages of minerals, the line between the "mine" and the rest of the planet became ever more tenuous.

Increasingly, the landscape of mass destruction was not just in Bingham or Butte or some other distant and isolated place—it was all around us.

We will return to Butte and Bingham in the pages to come, as the two constituted a sort of yin and yang in the development of modern mining: one "the richest hill on earth," the other "the richest hole on earth." Visiting these sites in their modern state suggests the source of the basic question at the heart of this book: how did the historical forces unleashed during the past century produce such radically scarred and transformed landscapes? The answer lies with Daniel Jackling and the invention of mass destruction technology during the twentieth century. The forces that created the Bingham and Berkeley terrestrial pits, however, began at an earlier and (literally) deeper level with the late nineteenth-century development of the immense subterrestrial hard-rock mines of the American West. There, mining engineers first learned to use powerful new technologies to overcome natural obstacles, pushing nearly a mile below the surface of the earth, where they created new human environments in which they had seemingly taken control of many of the basic elements of life: air, water, and climate. From such triumphant engineering creations arose the sometimes arrogant belief that the complex terrestrial environment could be as mastered as easily as the simpler subterrestrial environment.

Such grandiose dreams of power and control ultimately faltered, as both subterrestrial and terrestrial worlds proved more difficult to master than first believed. Indeed, even as engineers learned to create deep hard-rock mines, they had little concept of the immense galactic and geological forces that had placed the minerals within their grasp in the first place. In mining, human technologies and natural forces were inextricably linked from the start, all the way from the stars in the heavens down to the extraordinary physical properties of ancient copper atoms buried within the earth.

TWO

Between the Heavens
and the Earth

The operations of nature, mechanical and chemical, are supplemented
by those of man, who is a great mimic.

—T. A. Rickard

Thus the heavens and the earth were finished . . .

—Genesis 2:1

The autumn of 1902 in the Deer Lodge Valley was like many in southwestern Montana before and since. The days were mostly dry and warm, the
nights chilly and cloudless. Relatively little rain fell to wash the dust off the
quiet farming valley, and the hay and other crops grew predictably slower
as the daily hours of sunshine grew shorter. This was the typical cycle of
autumn, one that the ranchers and farmers had adapted to and learned to
plan for during the half century since the first Euro-Americans arrived in
the valley in the 1850s. What they did not know until much later was that
these welcome and predictable seasonal patterns were now conspiring
with poisons in the air to make their lands increasingly dangerous.

Deer Lodge rancher Nick Bielenberg was among the worst hit. He
watched with growing alarm and frustration as his once vigorous herds of
cattle, sheep, and horses suddenly began to sicken and die. The owner of
one of the oldest and most successful ranches in the valley, Bielenberg had
never seen anything like this. In the course of only a few weeks, more than

a thousand head of cattle, eight hundred sheep, and twenty horses were dead. When he talked with neighbors, he learned that many were experiencing similar stock deaths. In different circumstances, Bielenberg and the other ranchers might have blamed some exotic plague or the introduction of a new poisonous herb to the valley. In the autumn of 1902, though, a more obvious explanation lay on the far southwestern edge of the valley where the four tall stacks of the Anaconda Company's brand-new Washoe smelter sent a steady stream of stinking yellow smoke rolling out over the valley.[1]

So began the "smoke wars," the battle between the Deer Lodge Valley farmers and the Anaconda over whether the smelter smoke was damaging crops and livestock, and if so, what should be done to solve the problem. The Anaconda had built its first copper smelter in the sparsely populated Deer Lodge Valley in 1884, attracted to the area for the abundant fresh surface water available from Warm Springs Creek. The Washoe was fed by company ores from giant underground mines in nearby Butte, shipped to the smelter city over a dedicated rail line. The copper deposits were rich and astonishingly big, to an extent rivaled by only a handful of others in all of human history. The work of the Anaconda's corporate engineers had made much of this natural wealth available for exploitation, and the thousands of workers who labored underground pushed the mines deeper and wider every day.

By the time Nick Bielenberg's cattle began to sicken, collapse, and die, Montana had been a state for not much more than a decade. Nonetheless, this remote and sparsely settled region was home to one of the most advanced examples of modern mining and smelting technology in the world. The sheer size and scope of the Anaconda operations would present daunting challenges to the mining engineers, and their allied experts, as they attempted to understand and control the poisons killing Bielenberg's animals. Ultimately, their efforts would fail, though they made significant progress before the limits of their technology, a ruthless corporate drive for profit, and a ravenous American demand for copper undermined earlier successes. The result would be the dead zones, immense areas of environmental and human destruction. But that sad outcome was still several decades in the future.

Before then, the mining engineers approached the challenge of the Deer Lodge Valley smoke problem with optimism, confident that they could and must find a solution. To not mine, to abandon that colossal hill of copper ore, was simply not an option. In the minds of the engineers, as in the minds of most Americans, copper was the metal of modernity, the shiny red stuff that was helping to take the nation into a bright new age of prosperity and ease. After all, the man many believed to be the greatest inventor in the history of the world had made it so.

A BRIGHT LIGHT FROM A DYING STAR

The great breakthrough came on October 21, 1879, or so the popular story goes. That day one of Thomas Edison's Menlo Park laboratory assistants took a glass globe with a thin carbon filament suspended in a vacuum and connected it to the testing apparatus. With the close of a switch, electricity surged from a power generator, through a circuit of wire, and up into the bulb. It glowed—as was expected. Edison and his assistants had done this same experiment countless times before with platinum filaments. All of these earlier bulbs had also glowed, but the platinum filaments quickly burned out or gave only feebly flickering light. Edison's decision to try a new filament made by carbonizing a strand of common sewing thread was an act of creative desperation. Other inventors had tried carbon filaments before and failed. They, however, had not had Edison's advanced vacuum pumps, and the inventor hoped that a better vacuum inside the bulb would make all the difference. It did. The new bulb produced a steady glow of light, about as bright as the gaslight fixtures common then to American shops and homes. More important, the new filament burned nearly fourteen hours. After several months of work, Edison and his team found another type of carbon filament that burned for 1,200 hours. At last this was the long-lasting incandescent light bulb Edison had been searching for, the essential element in his plans to create an electrically powered lighting system cheap and reliable enough to replace gaslights.[2]

In reality, Edison's "eureka moment" was not quite so clear and dramatic. Testing of a carbon filament bulb stretched over several days, and Edison and his team only gradually realized they had found the answer. Historians of technology also know that Edison's "invention" was an important but only incremental improvement on the work of several generations of previous inventors. The carbon-filament vacuum bulb had been "under development," one might say, for nearly four decades by the time of Edison's 1879 experiments.

Most important, though, the traditional popular focus on Edison's light bulb tends to miss the significance of his true accomplishment: the creation of an entire system for generating and distributing electricity, without which the most advanced light bulb in the world was a useless brittle sphere of fancy glass and wire. As the historian Thomas Hughes argues, Edison's dynamos (electrical power generators) and his parallel-distribution wiring system were every bit as important as the light bulb itself.[3] Though rarely remarked upon by the public, both then and now, what was equally striking that day in October of 1879 was not just the light bulb but the fact that Edison was able to "plug it in." Of course, that first test bulb did not yet have the now familiar metallic screw base, whose brilliant simplicity would eventually spawn an entire genre of jokes. Rather, the bulb had two strands of insulated wire running down from the filament and through the neck of the glass containment vessel. To "plug in" the bulb, one of Edison's team secured the wires to a circuit that connected the bulb to the lab's newly completed high-voltage dynamo. Powered by a small coal-fired steam engine, the dynamo charged the wire circuit. When the researcher closed the switch, the circuit was completed, the electricity flowed, and the light bulb was able to glow for those fourteen revolutionary hours.

Although other metals like iron were in use as well, copper was critical throughout Edison's electrical power generating, distributing, and testing apparatus. The Menlo Park electrical system was really a small dress rehearsal for what was soon to come. Three years later, Edison installed his first electrical generating and distribution system on Pearl Street in lower Manhattan. A twenty-seven-ton "Jumbo" dynamo (named after P. T. Barnum's famous elephant) generated the electricity through a system of heavy copper bars and brass (an alloy of copper and zinc) discs that rotated around a magnetic core. Edison distributed 110-volt direct current power over several blocks via nearly twenty miles of thick copper wires threaded through underground conduits. Eventually, the Pearl Street Station supplied an area of about one square mile with enough power to light more than ten thousand bulbs. Customers liked Edison's soft, clean, and safe electric light, but the Pearl Street operation remained unprofitable for almost five years, in large part due to the immense cost of the heavy copper wire.[4]

It was fitting, then, that Edison sent the first-ever electric power bill, for $50.44 (a considerable sum at the time, equivalent to perhaps $1,000 in modern U.S. dollars), to the Ansonia Brass and Copper Company. A Connecticut manufacturer of copper and brass wire, tubing, and other materials, Ansonia had a New York office a few blocks from Edison's Pearl Street Station.[5] Ansonia had also provided Edison with some of the copper used in his Menlo Park experiments, and Edison had encouraged the company to improve its manufacturing methods in order to produce copper wire with high levels of uniform conductivity. In 1883, Edison provided Ansonia with a wire testing apparatus and the advice of one of his electrical experts, and he credited this for the company's development of a reliable means of making high-quality copper wire for the electrical industry.[6] With the Pearl Street Station up and running, the Ansonia managers surely understood that Edison's new electrical light and power system promised to create a major new source of demand for their copper products.

If electricity gave birth to the modern age, then copper was its mid-wife—or at least a very supportive birthing coach. Five years after Edison opened the Pearl Street Station, Frank Sprague built an electric streetcar system in Richmond, Virginia, creating a completely new consumer market for electricity. By 1902, the nation had 21,920 miles of electrified street-cars, most of them fed by copper wires and using copper-based electric motors. Manufacturers began to realize the benefits of electric lights and motors in earnest around the turn of the century, propelling demand beyond the consumer markets. With the development and eventual dominance of George Westinghouse's alternating current power (relegating Edison's prized direct current to a secondary role), centralized generating plants began using thick copper wires to transmit power over long distances. The growth cycle even fed back on itself as copper manufacturers used electrolytic refining to provide cheaper, purer, and more highly conductive products.[7] Between 1880 and 1914, the use of electricity in manufacturing went from essentially zero to almost nine million horsepower of installed equipment.[8]

Other markets for copper grew as well. By the mid-1920s, the Bell Telephone System had already bought more than seven hundred million pounds of copper for constructing its nationwide phone network.[9] Warfare also proved to be an especially voracious consumer. World War I sparked an enormous new demand, though it paled in comparison to the subsequent conflict. During World War II, eight hundred pounds of copper went into a typical tank, a ton into a large bomber, one thousand tons for a battleship. Brass shell casings demanded copper and zinc. A 37 mm antiaircraft gun could use a ton of copper during twenty minutes of sustained action. The machine guns of a fifty-plane squadron might shoot seven tons of copper in just sixty seconds of battle.[10]

Light and power, factories and streetcars, telephones and tanks—copper was essential to them all, not to mention the many new electrical devices like vacuum cleaners and blenders that were colonizing American homes during the first half of the twentieth century. Indeed, so many devices and processes depended on electricity during this period that per capita copper consumption became a fairly accurate indicator of national economic growth and modernization. Despite frequent turbulence in markets and a sharp drop in demand immediately after World War I, per capita copper consumption in the United States and Western Europe increased rapidly throughout the first decades of the twentieth century. Coal, iron, and steel may have been the building blocks of the world's first industrial age, but copper was at the electric heart of the second. Daniel Jackling, the inventor of the mass destruction technology that would give the world so much cheap copper, praised the metal that had made his fortune in a 1937 article. "Copper is verily one of the world's most essential metals," he writes. "Linked intimately with the production and application of electrical energy, copper plays a vital role in modern life. . . . Thomas Edison's incandescent lamp was the invention that started the tremendous expansion of that industry. With the widespread use of electric power have come most of our modern conveniences—the telephone, automobile, radio, refrigerator, washing machine, water heater and, lately, air-conditioning equipment."[11]

Many Americans also believed that copper would help the nation escape the grime and pollution that had plagued nineteenth-century industrialization. Copper would help usher in a clean, modern, electrically powered world that would provide all the conveniences of technology while also maintaining a healthy and morally uplifting environment. As Watson Davis notes in his 1924 hagiography of the metal, "If the cloud of smoke is to be lifted from our cities, if our factories are to be made clean and our homes convenient by the substitution of electric power for coal burning, it will be by aid of this humble element."[12] Even Lewis Mumford, who was at times a fierce critic of modern technology, initially praised the revolutionary potential of copper and electricity. In his classic 1934 history of Western technology, *Technics and Civilization*, Mumford argues electricity would help give birth to the "neotechnic" phase of civilization, an era with all the benefits of modernity but the "clear skies and the clean waters" of the preindustrial age.[13]

Mumford's sunny views stemmed in part from his belief that hydropower could meet much of the nation's demand for electricity. Edison's Pearl Street Station had used the dirty coal-fired steam engines of the past—what Mumford called the paleotechnic age. But many assumed the future of electricity would be hydropower facilities like those constructed in the late nineteenth century at Niagara Falls. There the diverted force of the river drove immense underground dynamos, and copper wires carried the power to distant cities. The modern miracle of hydropower turbines and long-distance transmission almost seemed to promise a belated return to the idyllic era of the pastoral water mill.[14] Copper, the mining engineer Charles Henry Janin enthused in 1924, was "taking the waterfall to the heart of the city; one instant tons of water drop; the next, tons of machinery hum."[15]

Copper mining companies often played important roles in the development of electric power generation and distribution, particularly in the West. Daniel Jackling helped found the Utah Power & Light Company, which initially provided power for the big new electric shovels in the Bingham Canyon pit but soon became Utah's main electric power company.[16] In Montana, the president of the Anaconda, John D. Ryan, founded the Montana Power Company that dominated electrical power generation and distribution in the state until the 1990s.[17] As a 1912 story in the *Salt Lake City Tribune* rightly observed, the Utah and Montana copper mining interests "paved the way for [copper], and by the development of their respective electric power fields they have made assured an adequate supply of power."[18] A wise move as it turns out, since by the late 1950s the Bingham Pit would be by far the largest single consumer of electric power in Utah. The pit shovels, trains, concentrators, and other machines consumed an astounding 650 million kilowatts of power every hour—roughly the amount used by a city of one hundred thousand.[19]

Americans used copper (and its common alloys, brass and bronze) in thousands of other ways, from Model T radiators to household plumbing. But it was copper's intimate pairing with electricity that drove much of the growth in demand during the first half of the twentieth century. That partnership hinged primarily on a seemingly trivial atomic quirk. Copper is a basic building block of the universe, one of the ninety-two atomic elements that occur naturally on earth (though some argue the number is slightly higher or lower) and from which all other substances are made. Its chemical symbol is Cu, an abbreviation of its ancient Roman name, *cuprum*. On the familiar chart of the Periodic Table of Elements that hangs in nearly every American chemistry classroom, copper appears as number 29. This atomic number helps to explain copper's unusually high ability to conduct electricity. The 29 indicates that each copper atom has twenty-nine positively charged protons in its nucleus and twenty-nine negatively charged electrons orbiting around the nucleus. Because of the way the electrons are packed in around the atomic nucleus, twenty-eight of them are in the lower electron "shells" closest to the nucleus. That means the twenty-ninth electron sits by itself in copper's outermost atomic shell, where it can be easily stripped off. This makes copper a wonderful source of moveable electrons. When electric current "flows" through a copper wire, the outermost electron jumps easily to the neighboring copper atom, which in turn shoves another electron down the line, and so on. The effect is something like pushing on one end of a long line of railcars. Each electron only moves an infinitesimally small distance, yet this line of colliding twenty-ninth electrons can transmit an electrical charge over very long distances almost instantaneously.

Because of its durability and resistance to corrosion, copper was also the metal of choice for steam pipes, plumbing, automobile radiators, and other heat exchange devices. It is very malleable, and manufacturers could easily form copper into a wide variety of shapes. Copper's high strength under tension also made it well suited for wire strung between power poles. By comparison, aluminum was lighter and cheaper but had only about 60 percent of the conductive capacity of copper and far less tensile strength—a serious shortcoming when a power wire sometimes had to span long distances between poles. Thicker aluminum wires could make up for the loss in conductivity, and the wires could be buried instead of strung from pole to pole, but the cost and disruption to built areas would have been much greater. Almost as if by design, copper seemed uniquely suited to serve the growing demands of national electrification.[20]

Immediately beneath copper in the same Periodic Table column are silver, number 47, and gold, number 79. Silver and gold are also superb electrical conductors for the same reason as copper: both metals have a single easily moveable electron in their outermost atomic shell, which is partially why they share a column on the Periodic Table (the "periodic" suggests these intriguing patterns of similarities in the elements that had long been remarked though not always fully understood). Edison did not use gold in his Menlo Park wires or silver in his Pearl Street Station dynamos for the obvious reason: gold and silver were expensive precious metals, whereas copper was a much cheaper "base" metal. To wire lower Manhattan with silver would have cost a fortune. To use gold would have probably sparked a bizarre New York "gold rush" with hordes of "eighty-twoers" covertly tearing up the pavement to mine Edison's precious wire. This distinction between precious and base metals, though, is substantially a function of relative scarcity and cultural fashions. If copper had been as rare as silver or gold, its culturally determined value might well have been as great or even greater. Compared to truly abundant mineral elements like aluminum and iron, copper actually is quite scarce. Geologists estimate that about 8 percent of the earth's crust is aluminum and 4 percent iron. Copper, by contrast, makes up only 0.01 percent of the earth's crust, or a concentration of about 50 parts per million (ppm). Even so, copper is still some seven hundred times more abundant than silver (0.07 ppm) and about forty-five thousand times more abundant than gold (0.0011 ppm).[21]

The crustal scarcity or abundance of all but the lightest elements (which scientists believe were created in the Big Bang some ten to twenty billion years ago) is in part a product of the stars, the original furnaces in which all the metallic elements were forged. Through a process called nucleosynthesis, the tremendous heat and pressure inside stars fused together the lighter elements to make heavier elements. Astronomers have long known that stars between one and eight times the mass of the sun gradually built up heavier elements in their cores over the course of millions of years. The heaviest of these is iron, and its relative abundance in the Milky Way (and on earth) is a product of this slow process of fusion in countless stars across the galaxy over the course of billions of years. Internal stellar fusion, though, can only create elements up to the atomic number of iron, 26. The nucleosynthesis of the remaining sixty-six elements that occur naturally on earth—including copper—thus must have resulted from even more extreme conditions than stellar atomic fusion.

Until very recently, scientists were uncertain how stars created the copper on earth. They knew that copper's close periodic cousins, gold and silver, were formed when massive dying stars blew themselves apart as spectacular supernovae. However, astronomers found that the distribution of copper in stars of different ages was not what it should have been if the metal was created by big supernova explosions. For years, astronomers debated the heavenly origins of copper. Some argued the metal must have been born in very big stars. Others were partisans of white dwarfs, dense "white-hot" stars smaller than the earth itself. Finally, a paper published in 2007 by two Italian astronomers appeared to have settled the debate. Most of the copper in the Milky Way, Donatella Romano and Francesca Matteucci conclude, arose in "supergiants," the most massive stars in the universe. Supergiant stars have ten to seventy times the mass of the sun, and they can be five hundred or even a thousand times bigger in size. As these supergiants consume the vast amounts of atomic fuel needed to keep them burning, they eventually reach a stage where neutrons are created. Unlike the supernova explosion with its immense sudden shower of neutrons, however, the supergiant core only produces neutrons very slowly. Copper, the astronomers argue, is created during this slow process of supergiant neutrons colliding with iron atoms and gradually heavier elements.[22]

Copper, gold, and silver are all the products of massive stars, then. Copper, however, was only forged over the course of millions of years, while gold and silver were created in the flash of an eye during the intensely bright explosive moment of a star's death. Since supernova explosions only occur in the Milky Way an average of about once every fifty years, we are unlikely to witness the stellar creation of gold and silver with the naked eye. But on any clear winter's night, residents of the northern hemisphere can easily see examples of the supergiant stars where more copper is even now slowly being created. Two of the brightest stars in the well-known constellation of Orion are supergiants: Rigel at the lower right of the constellation and marking one of Orion's feet, and Betelgeuse at the upper left and marking his shoulder.[23]

The vast majority of the copper, gold, silver, iron, and other heavy elements in the earth's crust today came from these stellar processes. When these elemental stellar nurseries exploded, they ejected their stores of metals out into space at immense speed, sending them on long cosmic journeys. Some of this material eventually gave rise to the next generation of stars and other solar systems, including the sun and its companions. After the formation of the sun itself consumed the mass of material, a thin cloud of dust remained in orbit around the young star. Over the course of millions of years, these leftover dregs from the sun's creation gradually accreted into the rocky bodies that became the planets, moons, and asteroids of our solar system. With the minor exception of occasional contributions from meteorites and other captured space detritus, the earth obtained all of the copper it would ever have when the planet formed out of this cloud of dust from shattered stars.

That copper is more abundant than the "precious" metals of gold and silver is thus a mere happenstance of stellar nucleosynthesis: like the tortoise, the slow but steady copper won out over the fleet but unreliable hare. As noted before, though, the earth's cosmic inheritance of copper is still relatively meager: only one one-hundredth of 1 percent of the planet's crust. Had this tiny amount of copper remained uniformly distributed throughout the earth's crust, it would have had a density of about 50 parts per million—much too small to be profitably extracted by even modern technologies. However, over the course of millions of years, geological forces concentrated a portion of the earth's total allotment of copper in a relatively small number of sites around the globe. On rare occasions, these deposits are pure elemental copper, so-called native copper. Far more commonly, copper is found in combination with iron, sulfur, or other elements.

Two of the most commercially important copper minerals are chalcopyrite and chalcocite. Because they both contain sulfur, geologists refer to these minerals as sulfides. Chalcopyrite ($CuFeS_2$) is copper iron sulfide, a molecule in which one atom of copper is bonded with an atom of iron and two atoms of sulfur. Chalcocite (Cu_2S) has no iron, so it is simply copper sulfide. Chalcopyrite was the most important copper mineral in the Bingham mines. Chalcocite was very abundant in the Butte mines, though other copper minerals were also present, as well as significant amounts of silver, gold, and other metals. The presence of such unusually high concentrations of minerals at these two particular sites was the result of two interconnected geological phenomena: plate tectonics and hydrothermal deposition.

The Butte and Bingham deposits were created millions of years ago when drifting tectonic plates of the floating crust of the earth collided. These forces slowly raised up the chain of mountains we call the Rockies, and in the process brought molten rock magma containing copper and other minerals much closer to the surface than is typical. As water from the surface percolated down into the earth, it dissolved copper minerals from the magma. This superheated mineral-laden water was then driven back toward the surface, where it gradually worked its way into cracks and fissures in the rock above. When the water cooled, chalcocite, chalcopyrite, and other minerals were left behind, but now in concentrations far higher than in the original magma. At Butte, where the rock itself was badly fractured, the deposition took the form of easily recognizable veins of minerals. Subsequent surface erosion and chemical weathering also leached out copper at the surface and redeposited it in a superenriched zone of chalcocite. Beginning at about three hundred feet below the surface of Butte and extending down to over a thousand feet, this immense deposit of high-grade copper ore was a major reason (though far from the only one) that miners called Butte "the richest hill on earth." The processes that formed the Bingham chalcopyrite deposits were similar, but with a critical difference, and will be discussed in detail in chapter 4.[24]

That Edison—and soon after, much of the nation and the world—could use copper to generate, create, and harness electric power was thus the result of an extraordinary chain of natural phenomena. In his groundbreaking work of environmental history, *Nature's Metropolis*, William

Cronon argues that Americans did not so much create the wealth they exploited in the West as simply take advantage of the natural wealth that was already there. Nature, in other words, had already done much of the work. While Cronon focuses on organic wealth from forests and prairie grasses, the same point holds true for the West's inorganic mineral wealth. The story in this book is mostly about the human efforts to win copper from the earth and control the resulting pollution. But whether we admire or condemn these efforts, we should not forget that humans were merely tapping into much larger galactic and geological forces of copper creation, dispersion, and deposition that had been going on for billions of years. Seen in this light, the mining engineers who developed the means to extract copper ore from Bingham and Butte were not nearly as different as they liked to believe from all the other animals that depended on the earth's natural bounty to survive and reproduce.

Most mining engineers, of course, did not view their technological achievements quite so modestly.

RATIONALIZING SUBTERRESTRIAL BUTTE

Walk the steeply sloping streets of uptown Butte, Montana, today and the seemingly solid ground beneath your feet is something of an illusion. Over the course of nearly a century, miners excavated an extraordinary ten thousand miles of mine tunnels and shafts, leaving behind something more like termite-infested wood than terra firma.[25] Sometimes the filigree of rock gives way and the ground collapses. The residents of Butte have long told stories of animals and buildings swallowed up by subsiding earth, disappearing into the ground as if pulled down by the very demons of hell. As the immense masses of rock below settled and shifted, pavement cracked and water lines splintered like toothpicks.[26] Though the days of underground mining have long since passed, ground subsidence remains a serious problem in Butte. Developers are reluctant to build in certain parts of the city where the ground can literally drop out from under their investment. Many of the more dramatic stories of ground collapse are fanciful exaggerations, examples of how the citizens of Butte take a perverse pride in emphasizing the jagged edges of their city. In an early scene from the 1971 biopic of motorcycle daredevil and Butte native Evel Knievel, the film tries to convey something of Knievel's hardscrabble youth. Walking in the barren dusty moonscape of Butte's mining district, the boy "Evel" blocks a car as he dawdles in the middle of a dirt road. When the ground suddenly collapses under the car and its driver, he listens with placid indifference as the blare of the car's horn slowly fades into the dark depths below.[27]

In Butte, the possibility that the subterrestrial will suddenly intrude on the terrestrial is always present. Usually, though, this underworld remains quietly invisible and largely unknown, except perhaps to the rapidly dwindling proportion of Butte's citizens who once actually worked in the underground mines. Merely being told of the reality beneath is of little help, as it remains difficult for "surface-lubbers" to envision or really even comprehend that the small city of Butte sits on the remains of an immense three-dimensional underground city that reaches more than a mile deep into the earth. Going into one of the few parts of the mines that are still accessible today is no help either. Being underground simply raises new barriers to seeing, since only a small section of the vast subterranean complex is visible at any one time. There are no "overlooks" for subterrestrial Butte, and no horizons or other fixed point of reference with which to get your bearings. Even the men who once spent many years laboring there only came to know a small part of the maze.

"Legibility," the political geographer James Scott reminds us, "is a condition of manipulation."[28] Any successful attempt to control and exploit a complex system, whether it be a subterrestrial mine or a terrestrial ecosystem, demands some means of taking its measure and mapping its essential characteristics. One of the greatest obstacles to large-scale industrial mining was developing the means for making the hidden underground world "visible." With improved maps and measurements, mining could be rationalized and systematized; obstacles could be overcome with powerful new machines and techniques. Gradually, the subterrestrial environment of the mine could be transformed into the physical embodiment of an "ideal" mine that had previously existed only in the heads of miners and mining engineers. During the late nineteenth and early twentieth centuries, American mining engineers did just that, creating an increasingly powerful array of techniques for measuring, engineering, and controlling complex underground spaces. As Rossiter Raymond, the secretary of the newly created American Institute of Mining Engineers, said in 1871, the goals of the profession were clear: "We want analysis; we want measurement; we want exact comparison; we want the universal recognition of the absolute value of the truth, and the relative worthlessness of anything short of it."[29]

In this adamant demand for analysis, measurement, and what Raymond considered "the truth" lay the seeds of both the mining engineers' modern success in creating and controlling gigantic but comparatively simple subterrestrial environments and their failures in managing more complex terrestrial environments. The extreme expression of these ideas and techniques would be Daniel Jackling's open-pit mine at Bingham, a monument to the way in which modern rationality could sometimes become profoundly irrational.

Yet for all its power, the engineers' control of their underground environments was also flawed. Since at least the days of the ancient Romans, mine operators had pushed the limits of their technical abilities in order to make a profit from extracting minerals at ever greater depths. As mines began to sink thousands of feet below the surface, the problems of subterranean flooding, heat, and ventilation grew ever more challenging. The

4. Early twentieth-century miners drilling in the highly engineered environment 1,900 feet below the surface in Butte, Montana. As the Butte mines sank deeper, mining engineers created complex pumping, ventilation, and cooling systems that made it possible for humans to survive and work in harsh underground environments. *Courtesy Montana Historical Society Photograph Archives, Helena.*

mine engineer was increasingly required to become a type of early environmental engineer, developing complex technological and managerial systems to meet the basic biological needs of the human beings who actually worked in the deep underground. As with the mines themselves, these early life support systems were an extraordinary engineering achievement, and they eventually permitted miners to survive and work in the harsh environments more than a mile beneath the earth. However, when complex envirotechnical systems failed, the consequences could also be disastrous. In this, the triumphs and the disasters of engineering mining ecosystems anticipated those that would meet later attempts to engineer surface ecosystems.

Though the United States had long engaged in coal mining, and miners had developed underground lead deposits in Illinois and Missouri, large-scale hard-rock mining operations did not become common until after the California Gold Rush of 1849. Of course, the early California gold miners came in search of easily obtained placer deposits, not hard-rock lode deposits. These rich deposits of gold dust and nuggets were found along the courses of rivers and creeks, and they quickly played out. Most of the unskilled miners gave up or headed elsewhere in search of the next big strike. In the decades to come, Nevada, Colorado, Montana, and other western territories would all have their brief moments as the next big placer bonanza. A few of the more ambitious California miners managed to extend the life of the placer mines through hydraulic mining, a powerful early mass-mining technique the Romans had used on Spanish placer deposits two millennia before. Hydraulic miners dammed a creek or river to build up a head of water pressure and then channeled the water down through a heavy fabric hose to a cast-iron cannon, or "monitor." As the water entered the narrow barrel of the monitor, it built up enough pressure to shoot out of the nozzle at speeds of over one hundred miles per hour. Miners scoured away entire hillsides of dirt and gravel with the high-speed jets of water, capturing the relatively small amounts of placer gold by channeling the runoff down long sluices.[30]

As the historian Andrew C. Isenberg notes in his pathbreaking ecological history of California industrial mining, the advent of hydraulic mining there signaled a shift to the capital-intensive industrial mining that would dominate much of the West for the next century. Whereas the forty-niners had used simple tools to develop placer mines, the hydraulic miners used costly tools and large-scale engineering techniques. Only the few who had access to scarce outside capital could afford to develop such mines, and increasingly the men who worked them were wage earners rather than owners.[31] The powerful but imprecise methods of hydraulic mining also exacted a heavy toll on the California environment, washing away thousands of acres of forests and soil. The silt generated by the mines choked downstream rivers, destroyed fisheries, and covered good farmland with thick layers of unproductive sand and gravel. As Isenberg rightly argues, hydraulic miners shifted much of the cost of industrial mining to the environment, as would be the case with the subsequent development of industrial logging, ranching, and farming in California.[32]

This practice of achieving industrial speed and efficiency by transferring costs to the environment would reach its extreme expression half a century later in the mass destruction techniques pioneered by Daniel Jackling at his Bingham Pit. Jackling's nonselective copper mining technique bears a familial resemblance to hydraulic gold mining. However, the engineering challenges to open-pit low-grade copper mining were far greater than those posed by hydraulic mining. Thus, a more significant technological precursor to Jackling's pits was the development of deep hard-rock mining operations. Most placer miners had neither the desire nor the ability to find and extract the gold from its original source in the Sierra Nevada, the so-called mother lode. But as with the hydraulic miners, a handful pioneered the early efforts at hard-rock mining. The most significant of these early mines was the Comstock Lode, an immense aggregation of silver and gold deposits on the eastern slopes of the Sierra Nevada near Lake Tahoe. To mine the Comstock deposits demanded a quantum leap in technology and capital well beyond even the most ambitious hydraulic mining operations.

Unfortunately, few Americans in the mid-nineteenth century had any experience in hard-rock mining. The most successful mine owners and operators sought the aid of European miners and mining engineers, many of whom had learned their craft in the mines of Europe and England. Such was the case with the early Comstock miner George Hearst, a man whose own considerable fame was later eclipsed by that of his son, the newspaper baron William Randolph Hearst. Orson Welles drew loosely on William Hearst's life in writing his 1941 film *Citizen Kane*. Welles offered a Colorado gold mine as the basis of the Kane family fortune, which seemed an obvious allusion to the Nevada silver mine that was the source of Hearst's. Like many California miners, Hearst initially tried his luck with the placer gold mines and found little success. In contrast to the vast majority of the Gold Rush miners, though, Hearst knew something about lode mining, having worked in the lead mines near his Missouri home. When he heard reports of a rich gold and silver lode deposit on the eastern flanks of the Sierra Nevada, Hearst rushed to the site and established several claims. Joined by his financial partners, James Ben Ali Haggin and Lloyd Tevis, Hearst began development of the Ophir mine on the Comstock Lode in 1859.[33]

Technical obstacles quickly emerged. At only fifty feet below the surface, Hearst's miners hit groundwater and the mine began to flood. Fortunately, Hearst and his partners had the capital to purchase the first steam-powered pump to dewater the Comstock Lode, and they then paid a good deal more to have it hauled by wagon from San Francisco. An eight-inch pump capable of generating forty-five horsepower, the machine was the direct descendant of the early rocking beam steam engines pioneered in the Cornish copper and tin mines of England over the previous century. At 175 feet, Hearst and his miners faced an entirely new problem. Despite its being a "hard rock" mine, the ore and surrounding country rock on the Comstock Lode were actually exceedingly soft. Miners could gouge out the silver ore from the stope with nothing more than a sharp pick, which made extraction an easy but dangerous proposition. As the mine went deeper, the sheer weight of the rock above would crush conventional wooden timber props, and the entire mine was in danger of collapsing. The solution came from another European import, this time in the person of Philip Deidesheimer, a German graduate of the best mining school in the world, the Bergakademie (School of Mines) in Freiberg, Saxony. Deidesheimer developed a new "square set" timbering method that used heavy interlocking wooden beams to create a series of hollow cubes that could be extended indefinitely upward and to the sides. The process was somewhat analogous to constructing the skeleton of an immense wooden building—one that was thousands of feet tall and wide—entirely underground.[34]

Over the next two decades, Hearst's Ophir and the neighboring Comstock mines went deeper, some of them eventually reaching down to three thousand feet. To one contemporary visitor, the mines were like an underground industrial complex, a subterrestrial "city 3 miles long and half a mile wide." Others thought it resembled a huge series of interconnected tenement buildings, a sort of underground sweatshop for the mass extraction of silver ore.[35] Indeed, the reference to sweatshops was more than mere metaphor, as the air temperature in the mines increased by around three degrees with every hundred feet in depth. In the deeper zones, ambient temperatures reached 110 degrees Fahrenheit, while the groundwater percolating up from below was a scalding 170 degrees. The Comstock mine engineers improved ventilation by driving long horizontal adits from the side of the mountain into the mine workings, helping to encourage the passive circulation of air. Eventually, though, engineers had to install the largest pumps and blowers then available in the United States to transport surface air into the depths. Nonetheless, parts of the mines remained unbearably hot. Some companies lowered barrels of ice water into the mines. One Comstock miner wrote in his diary that work in the hot mines was still so arduous that it made him physically ill. Ironically, he was later killed when his sleeve caught in a ventilation blower that brought cool surface air down into the mine. Rather than having his arm cut off for quick transport to the surface for treatment, he chose to remain below in hopes of being extracted in one piece. He bled to death while he waited.[36]

The growing human cost of deep-level hard-rock mining would become even more evident in the highly engineered underground copper mines of Butte. Like the Comstock, Butte began as a placer gold mining camp before evolving into a gold and silver lode mining operation. Initial capital for the hard-rock lode mines came from local entrepreneurs like William Andrews Clark, a Montana merchant and banker who had shrewdly profited from earlier gold rushes in the territory. Recognizing the necessity of technical knowledge in successful hard-rock mining and smelting, Clark even traveled east to the new Columbia School of Mines in New York (an institution modeled on the famous Freiberg Bergakademie) for a semester of studies in 1872.[37]

Four years later, further technical expertise and the promise of big outside capital came to Butte with the arrival of Marcus Daly, a representative of two Salt Lake City smelter operators intrigued by the possibilities of Butte's silver ores. The son of impoverished Irish potato farmers, Daly immigrated to the United States in 1856 at the age of fifteen. He followed the mining frontier to California and eventually found his way to the Comstock Lode. Smart, gregarious, and affable, Daly rose quickly in the ranks of the Comstock miners, learning all he could about the technology of deep lode mining at the Comstock. Perhaps even more important for his future career, Daly won the trust and respect of George Hearst, the San Francisco capitalist who had made his fortune with the Ophir mine.[38]

Daly put his mining experience to good use in Butte, taking an option on a small but promising silver mine called the Anaconda that was a mere sixty feet deep at the time. He convinced his old Comstock friend, George

Hearst, to invest heavily in the mine, and Hearst brought in his partners, James Haggin and Lloyd Tevis. As the Anaconda went deeper, however, the signs grew that the mine contained more copper than silver. In late 1882, at a depth of about three hundred feet, Daly's miners hit an extraordinarily rich vein of the copper sulfide ore chalcocite. Daly asked Hearst and his partners to invest even more capital with the goal of creating a major copper mining and smelting operation on the isolated Montana mining frontier. According to the traditional story, Hearst and Tevis initially hesitated. Neither they nor Daly had any experience in copper mining, and it was not at all clear that a base metal like copper could be profitably mined so far from the nation's urban centers. Haggin, however, argued that Daly's judgment had always been keen in the past, and he convinced Hearst and Tevis to invest millions in the Anaconda operation.[39] A year later, the mine was six hundred feet deep and the rich copper vein was more than a hundred feet wide—one of the biggest copper deposits ever discovered.[40]

Where the Comstock had previously pioneered deep-level mining, the Anaconda now increasingly took the lead. Though he was not a formally trained mining engineer, Daly had learned a good deal about industrial mining through his apprenticeship on the Comstock. There he had the chance to inspect elaborate steam-powered water pumps, lifts, and ventilation systems. Unlike some western mine managers who had learned their trade on the job, Daly also valued the abilities of college-trained engineers, geologists, and other experts.[41] The contributions of the Freiberg-trained Phillip Deidesheimer to mining on the Comstock would not have escaped his attention. Perhaps his favorable impression of Deidesheimer later influenced his 1891 decision to hire August Christian as his chief engineer, another German native who had graduated from the Freiberg Bergakademie. Christian had overall authority over the engineering operations at the Anaconda mines until his death in 1914.[42]

Daly and his successors at the Anaconda also increasingly relied on the expertise of academically trained surveyors and geologists. Aside from rough sketches of developed mine works, elaborate underground maps were rare in the American West at the time. In a small mine with limited investment in hoists, pumps, ventilation systems, and mills, it might suffice for a mine manager to simply follow surface signs of ore down into the earth, coming to understand the shape and structure of the deposit only as the excavation of the shafts and tunnels slowly revealed it. For a deep and sprawling mine like Daly's Anaconda, though, it became increasingly important that the engineer and mine managers develop as much knowledge as possible about the extent and nature of the deposit before mining began. Investors providing the big capital essential to developing the mine and mill wanted some reassurance that there was enough valuable ore to justify their investments. Absent good maps of the underground geology of Butte, managers like Daly had little choice but to do a great deal of initial "dead work"—shafts and tunnels that probed the size and richness of deposits but produced little profitable ore. Dead work was a costly means of envisioning subterranean space, but it was preferable to blindly following an ore seam into the ground where it might suddenly disappear.

Creating and efficiently managing deep mining operations demanded new ways of seeing and managing underground space. Daly had a reputation for being able to "see further into the ground than any other man."[43] But there were limits to what even a savvy and experienced mine manager could do without more advanced technical and scientific tools. Fortunately, for Daly and the other Anaconda managers who came after him, they were able to draw on the tools of another emerging profession, the mining geologist. As the geographer Steven Braun notes in his intriguing recent history of the development of "verticality" in Canada, by the end of the nineteenth century geologists, mining engineers, and governmental mining personnel had given "depth" to territories that had previously been seen in strictly terrestrial and (typically) agricultural terms.[44] In contrast to farmers, ranchers, or others with primarily agricultural interests, one of the striking aspects of the mining geologist's metric of visualization was the way it often ignored the traditional terrestrial human environment. Surface topography, rivers, vegetation, structures—many terrestrial features were ignored in order to better clarify the structure of the subterrestrial. In this narrowing of focus, of course, lay the visual power of the geological map. But it is also important to bear in mind how such maps helped to construct the subterrestrial environment as distinct from and unrelated to the terrestrial environment. Such a view would have its ultimate expression in Daniel Jackling's open-pit mine where the surface was literally stripped away, just as it had been symbolically stripped away in the mining geologist's map.

The great importance the Anaconda assigned to precise geological knowledge and mapping was ultimately reflected in the company's 1924 decision to combine the previously separate Geological Department with the Mining Engineering Department, both now under the direction of the "Chief Geologist and Engineer."[45] The increasing role of the geological sciences and mapping, however, should not obscure the continuing importance of the mining engineer's work in actually constructing the underground spaces necessary to exploit this knowledge. Mining engineers and managers not only adopted the geologist's construction of the underground world as an idea; they took on the challenging task of transforming that idea into an actual physical space, an environment in which humans could, however imperfectly, survive and work.

The Anaconda managers and engineers faced a multitude of challenges to carving out underground spaces, including the age-old problems of too much water and not enough air. The miners in Butte first began to intersect groundwater flows in 1877, discovering that the subterrestrial environment was home to vast lakes and rivers. The major groundwater bodies appeared at about 140 feet below the surface. Marcus Daly ordered two gigantic pumps from the Knowles Steam Pump Works of Warren, Massachusetts. Only a year earlier, the *Manufacturer and Builder*—a self-proclaimed "practical journal of industrial progress"—had praised the Knowles pump as "the most powerful and efficient steam pump ever offered in the American market." The big pumps were ideal for draining water in deep hard-rock mines, the journal suggested: "The Knowles Steam-Pump Works built pumps guaranteed to pump water from mines from 100 to 1,000 feet in depth," and the pumps could handle the gritty and corrosive water of mines without frequent repairs.[46]

Perhaps Daly read this very article when contemplating how to solve his flooding problems in the Anaconda, though he would likely have known about the famous Knowles pumps already. The pumps were expensive—Daly had to pay $100,000 for two of them, roughly the equivalent of at least $2 million today. To continue mining below the groundwater table, though, Daly had to have these sophisticated machines. Butte miners without the necessary capital or technical expertise to continually pump water up from the depths and out into the terrestrial environment had to either sell out or shut down. For the next century, mining in Butte essentially took place not only in an underground space, but in what had been an *underwater* space—one now kept dry only by the constant efforts of gigantic subsurface pumps.[47]

The pumps allowed Daly and other Butte miners to continue deeper into the earth, but this created new problems as well. As was the case at the Comstock, the Anaconda mines grew hotter with depth, and the heat and groundwater made for humid conditions. Bad or insufficient air was an old problem for miners, but the subsurface air quality problems at Butte and other mines were further aggravated in the late nineteenth century by the replacement of human-powered hand drilling with steam-powered pneumatic rock drills. The rapid hammering action of the drills produced clouds of fine silica dust that lodged in the lungs of the miners. Many succumbed to silicosis as a result, a deadly lung disease all too common in Butte and many other western hard-rock mining towns.[48] In 1918, the Anaconda installed one of the most elaborate mining ventilation systems in the nation, using immense electric fans and miles of flexible canvas tubing to force surface air deep down into the shafts and through the mines. At the Mountain Con mine, which was over four thousand feet deep by the 1930s, mining engineers even built an elaborate evaporative cooling system in order to make a tolerable (though far from pleasant) environment for human labor.

The Butte mines also pioneered the use of technologies that permitted humans to survive and work in poisonous subterrestrial atmospheres. In 1853, a Belgian university professor developed a device to provide an artificial supply of oxygen for use by miners in noxious underground environments.[49] This "oxygen breathing apparatus" fed a fixed stream of oxygen from a pressurized tank to a head mask, while the wearer's exhaled breath (which still contained considerable oxygen) was recycled through a container of lye to absorb the carbon dioxide. Other devices based on similar principles followed during the next half century, exchanging technical improvements with the parallel development of deep-sea diving devices.[50] H. A. Fleuss developed and marketed a machine in London as early as 1880, and in 1903 Bernhard Draeger of Lübeck, Germany, began selling a popular device. Initially, all these breathing machines were able to provide only a limited supply of oxygen, and many of them were awkward, flimsily built affairs poorly suited to the rough treatment they had to face in a mine.[51] Still, for decades they offered the only means by which humans could enter into mines where the air was poisonous or lacked adequate oxygen.

The first subterrestrial breathing machine arrived in Butte in 1907. The Boston & Montana Smelting & Refining Company (a company later absorbed by the growing Anaconda combine) purchased five of the machines from the Draegerwerk in Lübeck. The Anaconda soon after bought several Draeger units as well.[52] The device used two steel cylinders charged with oxygen at 120 atmospheres, which was enough to last two hours in ideal conditions, though the ideal was rarely the case in practice. The oxygen from the tanks passed through rubber tubes to either a helmet tightly sealed around the face or a valve held directly in the mouth. The helmet model—an often unnecessary holdover from the deep-sea diving use of similar machines, and one that led the wearers to be dubbed "helmet men" —had a sponge wired inside to wipe away moisture that formed on the mica face shield. Condensation problems were avoided with the mouthpiece model, in which case the wearer's nose was held tightly shut with a clip. However, the helmet design did keep thick smoke and poison gases away from the eyes. As the more efficient mouth-fed devices gained in popularity, miners wore goggles rather than the bulky and constricting Draeger helmets.[53]

Within months of taking delivery of its first five Draeger machines, the Boston & Montana put them to use when a fire broke out in the company's Minnie Healy mine. Despite the deadly streams of dense smoke choking the subterrestrial mine passages, miners wearing the Draeger devices went down the shaft, made their way through the tunnels, and built several sturdy bulkheads to contain the fire. They returned to the surface unharmed, having saved the Boston & Montana thousands of dollars in damages and ore losses. A few months later, the U.S. Geological Survey (USGS) tested its new Draegers after a horrific explosion at the Monongah coal mine in West Virginia that took 356 lives. Donning a Draeger helmet, a USGS investigator was able to immediately descend straight to the explosion's center while the area was still devoid of oxygen. With the evidence fresh, he was better able to determine the cause of the explosion.[54]

For the first time in the long human history of mining, miners could actually carry their own supply of air. Doing so was still dangerous, though, and miners needed careful training to properly operate the machines. Just as important, the miners had to learn to suppress the sense of panic and fear that often seized inexperienced users when they entered a potentially deadly mine. Miners were taught to trust the life-sustaining power of the machines they carried on their backs—even though that trust was not entirely warranted. A 1923 Bureau of Mines study found that twenty-seven "helmet men" had died that year while conducting rescues or doing other work in a toxic mine environment. This was a fatality rate of 1.2 percent. The report did not list the number of men who were seriously injured, but it must have been much larger.[55] Despite the risks, more than seventy thousand miners took the bureau training course between 1911 and 1940, and they undoubtedly helped to save the lives of many of their coworkers.[56]

Some companies also demanded that miners use the machine for considerably less noble reasons. Many took advantage of the technology to speed repairs in damaged mines, or even to resume mining in environments that would otherwise have been deadly. In 1911, a huge underground fire at the Phelps Dodge Copper Queen mine in Arizona inundated some areas of the mine with highly corrosive sulfur dioxide smoke. At depths between a thousand and thirteen hundred feet, temperatures hovered around ninety-seven degrees with nearly 100 percent humidity. This sulfuric acid steam bath ate away at the mine hoist cables and the bolts that held the hoist cage together. To keep the hoist operating and the mine open while the fire burned, Phelps Dodge sent men wearing Draeger oxygen helmets into the shaft day and night to repair the machinery. In other cases, companies demanded that miners use the helmets merely so they could continue extracting ore in a toxic section of the mine.[57] In an earlier age, these mines would have been abandoned, but with breathing machines, miners could continue to work in such dangerous underground environments.

Mining engineers and managers at the Anaconda and other big copper mines thus repeatedly used technology to surpass the subterrestrial environmental barriers to human survival. With these technologies, engineers, managers, and miners created a wholly new type of human environment, one in which the seemingly solid categories of the "natural" and the "artificial" became confused and intermixed. Groundwater removed from the mines by steam pumps became part of natural surface water drainages. "Natural" surface air, artificially cooled and transported by giant fans and hoses, became part of the subterrestrial environment where miners breathed and worked. The miners themselves moved continually between terrestrial and subterrestrial worlds, their hybrid machine-dependent lives underground coming to seem as natural as their terrestrial lives under the sun. With Draeger machines, men could even survive for a few hours' time in subterrestrial atmospheres that would otherwise have quickly killed them.

It is difficult to find an earlier or more striking example of the emerging modern ability of human beings to create habitable environments in hostile natural circumstances. In many ways, these deep subterrestrial environments were the precursors to the modern engineered environments in jets and spacecrafts that sustain human life in equally hostile circumstances. However, the degree of the mining engineers' ability to create and control subterrestrial environments should not be overstated. Despite the engineers' best efforts, the mines remained challenging, complex, and dangerous places. Further, the engineers also continued to depend on the expert knowledge of the underground workers to carry out the actual mining. The miners' knowledge of the layout of the mine, their ability to judge the strength of the rock and whether it needed timbering, and their skill in using explosives with maximum effectiveness were all essential to

5. Operators pose proudly by one of the big steam-powered lift engines that made deep-level mining in Butte possible. With such powerful hydrocarbon-fueled machines, engineers were able to create hybrid technological and natural subterrestrial environments, early precursors to the life-support systems found in modern jets and spacecrafts. *Courtesy Montana Historical Society Photograph Archives, Helena.*

the efficient production of the mine. Control of the mine also did not lead to the elimination of mining accidents and industrial diseases like silicosis. To the contrary, the rationalization of the mine for increased speed and production could often create new safety hazards.[58] In decades to come, worker safety became an increasingly important goal for the Anaconda and other big mining companies, but the mines remained very dangerous environments for workers, as do hard-rock mining operations to this day.[59]

As many historians have pointed out, the continuing dangers of underground played a role in the formation of unions among the western miners. It was no coincidence that the technologically advanced mining city of

Butte was also an important early center of organized labor. For a time, Butte well deserved its nickname "the Gibraltar of unionism." The Butte Miners Union was one of the earliest and most powerful unions to form in the western hard-rock mining districts, and it played a major role in the later creation of the nationwide Western Federation of Miners. Thanks in part to union efforts, Butte's miners were among the highest-paid workers in the nation at the turn of the century.[60] The Butte unions weakened in the early twentieth century, however, from a combination of hostile management attacks and splits among the workers themselves. The technological rationalization of the subterrestrial environment that helped increase and maintain productivity also undermined the power of the miners, as knowledge and control increasingly shifted into the hands of the engineers and management.[61] As the Butte miners became more dependent on complex technologies—not only to do their work but also simply to stay alive—the relative independence they had once enjoyed underground diminished. Both as workers and as living and breathing biological organisms, the miners became ever more deeply enmeshed in and dependent upon Butte's vast subterrestrial envirotechnical system.

When these elaborate life-support systems failed in some minor way, as they often did, the consequences were not always serious. Water pumps, ventilation systems, hoists, and all the many other pieces of the machinery of underground life occasionally broke or malfunctioned. Managers would dispatch mechanics and engineers to repair the problem, and such breakdowns were mostly annoying inconveniences to the working miners. But when these complex systems failed disastrously, the consequences could be deadly. The worst hard-rock mining disaster in American history occurred at Butte's Speculator mine, a property owned by one of the few remaining operators not yet controlled by the Anaconda. On the night of June 8, 1917, workers were lowering an electrical cable twelve hundred feet long and five inches thick down the mine shaft. Weighing nearly three tons, the cable—made of a highly purified form of the very copper minerals that other miners were removing from the Speculator at that moment—was needed in order to complete installation of a new sprinkler system designed to prevent fires in the mine. While lowering the cable, though, the workers accidentally dropped it down the shaft and subsequently set fire to the oil-soaked insulation wrapped around the copper wire. The fire quickly spread up the length of the cable and then jumped to the wooden planks lining the shaft itself. Within minutes, the shaft had become "like a gigantic torch," as one horrified Butte resident later recalled.[62]

In the hours and days to come, the fire in the mine would eventually claim the lives of 164 men. Only a few died in the flames. Most were killed by the invisible and odorless carbon monoxide gas that stealthily spread throughout the underground tunnels. The gas killed some men before they even knew what had happened. Others died only after long desperate hours waiting in the darkness for rescue crews that came too late. Rescuers did manage to save many men who otherwise would have likely perished, thanks in large part to the breathing machines that allowed them to enter the still poisonous mine passages.[63]

In the weeks after the Speculator mine disaster, workers responded to the tragedy with strikes and the formation of a new union. Although the deaths of 164 men in the massive fire clearly played a role in reviving union interest, demands for improved safety and working conditions were secondary to calls for union recognition and better wages.[64] Indeed, by the standards of the day, the North Butte Mining Company that operated the Speculator was already widely viewed as one of the more safety-conscious companies in Butte.[65] For some miners, though, the catastrophic failure of the company's subterrestrial life-support and safety systems was symptomatic of much deeper failings in the entire system of industrial mining. As one of the union handbills distributed after the fire argued, the "terrific holocaust at the Speculator mine" had its root cause in management's insistent demands to maximize ore production and "GET THE ROCK IN THE BOX."[66] Anger over corporate failings to engineer a reasonably safe working environment could easily shade into more fundamental anger over the injustices of corporate capitalism itself. In the turbulent wake of the disaster, the staunchly anticapitalist International Workers of the World (IWW) gained some ground among the Butte miners. The Anaconda responded by importing two hundred hired "detectives" and spies to infiltrate and undermine all union efforts, radical or not. The Anaconda's "goon squads" may well have played a role in the grisly murder of IWW organizer Frank Little, though the historical record is equivocal on this point. The IWW organizer had more than a few enemies, and his killers might have been rival union men or hyperpatriotic Montanans angered by his vocal criticism of American involvement in World War I.[67] Either way, few tears were likely shed among the Anaconda managers upon hearing of Little's death.

Given this volatile mixture of death and murder, anger and retribution, unions and capitalism, it can be difficult to place the Speculator mine disaster in its broader context as an industrial accident. But whatever its meaning as a symbol of corporate greed or the failings of capitalism, the Speculator fire must also be understood as a massive failure of a complex technological system. In retrospect, the fire appears to have been a case of what the organizational theorist Charles Perrow calls a "normal accident." Perrow argues that normal accidents inevitably occur in highly complex and tightly interconnected technological systems. Within such systems, even relatively small failures can cause an unforeseen chain of effects that lead to disaster. Even efforts to engineer in safety and fail-safe measures can inadvertently contribute to a catastrophic system failure. Since engineers and designers cannot possibly predict the course of these highly dynamic failures, Perrow argues, accidents will be an unavoidable, if rare, consequence of the normal everyday operation of such systems.[68]

In this sense, the Speculator disaster appears to have been at least in part a consequence of the growing complexity of the subterrestrial envirotechnical system. As Perrow's theory suggests, even the engineering efforts to increase the safety of the mine contributed to the tragedy. As already noted, the men were lowering the electrical cable that day as part of a project to install fire-suppression sprinklers in the shaft. The cable fed electricity to a transformer, which in turn provided power for electric lights and other devices that significantly improved mine working conditions (though they also created the new danger of electrical shocks). The oil-soaked cable insulation that caught fire was also a safety measure, a means of protecting workers from electrical shocks in the era before plastic insulation. That the small fire had spread so quickly and made the shaft into a "gigantic torch" was partially the unintended consequence of engineering efforts to improve the underground atmosphere. The Speculator had a reputation for being one of the best-ventilated mines in Butte thanks to the brisk circulation of air down one shaft, though the mine tunnels, and out a second shaft. The night of the disaster, this ventilation system that normally brought vital air to workers both fanned the burning flames and carried the resulting smoke and carbon monoxide deep into the most distant mine passages. Tragically, an envirotechnical system designed to bring cool fresh air to the miners now brought them carbon monoxide and death.[69]

Of course, none of this is to suggest that the managers and engineers of the Speculator could not have done more to make the mine safe. They certainly could have, and subsequent investigations suggested at least some level of company negligence as well as abundant evidence of missed opportunities to improve safety. At another level, though, the Speculator mine disaster was a product of the growing complexity and interconnectedness of early twentieth-century underground mining. Engineering and managing deep subterrestrial environments like those at Butte were becoming more complex and difficult, particularly in terms of the environmental support systems that kept workers alive at ever more dangerous depths. Problems that might have been easily dealt with in a two-hundred-foot-deep mine, or perhaps would have never occurred at all, had far graver consequences in a three-thousand-foot-deep mine—especially for the vulnerable human beings whose survival depended on machines. Thus, just as hydraulic mining and open-pit mining shifted some of the costs of efficient development to the natural world, so too did deep underground mining shift some of the costs to the humans who worked there. Even technologies designed to sustain subterranean human biology could at times prove deadly.

For more than two thousand years, the human drive to push ever deeper into the earth in pursuit of minerals has created growing challenges and dangers. The Romans, careless of the lives of slave workers, used only the most basic technologies in their mines of Rio Tinto, creating subterrestrial hells a thousand feet below the surface. Saxon miners of medieval Germany reached similar depths but created somewhat more humane working environments with ingenious water-powered pumps and fans. The British and later the Americans used steam engines and the concentrated energy of coal to increase their power to dewater and ventilate mines, making it possible for an industrialized workforce to labor in the extreme heat and humidity two-thousand feet down in mines like the Comstock Lode. Coal, and later oil and hydropower, sustained human life and work at four thousand feet beneath the city of Butte by the 1930s, providing the essential energy to drive immense evaporative air conditioners and fans on the surface. Thirty years later, the engineers of Anaconda's Mountain Con mine used similar technologies to push more than a mile below the surface and earn the city its motto: A mile high, a mile deep.

The engineering of these immense subterranean human worlds was one of the greatest technological achievements of the modern age. However, the hidden mines of Butte have never earned the mixture of attention, admiration, and criticism lavished on more obvious terrestrial icons of modernity like the Empire State Building, the Panama Canal, or Hoover Dam. Imagine, though, if the mines of Butte were somehow carved out of the surrounding rock as a single block, lifted up, inverted, and set back down on the surface. The resulting structure of stone, steel, and wood would, in many regards, far surpass any other on the planet. A mile high at its tallest point, the Butte mines would be twice as big as the world's largest skyscrapers, while the 726-foot-high Hoover Dam would be dwarfed in comparison. Nearly two miles thick at the base, the mines could easily contain dozens of Pentagon-sized buildings within just their lowest levels. Likewise, the Pentagon's much ballyhooed 17.5 miles of corridors are minuscule in comparison to the mines' 10,000 miles of vertical and horizontal passages. Of course, a roughly cut mine passage is hardly the same thing as a Pentagon hallway or a corner office on the ninety-first floor of the Empire State Building. On the other hand, though, the engineers of those structures did not have to provide for the constant removal of water or deal with ambient air temperatures over one hundred degrees Fahrenheit.

The point of such comparisons is not to minimize these other engineering accomplishments so much as to highlight how the extraordinary achievements of modern mining engineering have often been overlooked—in part, simply because their creations are hidden underground. The oversight is all the more unjustified when the mines are seen in their proper light as early examples of highly sophisticated environmental engineering. In what other environment had engineers achieved an equally clear expression of the high modernist drive to rationalize, simplify, and control nature? In the mine, the factory was embodied in nature, the natural and technological surrounded and enmeshed the human, and the human worked in a hybrid world where technology and nature were inextricably fused. In *Seeing Like a State*, James Scott offers the rationally planned German forests of the nineteenth century and the futuristic cities of Le Corbusier as prime examples of the material expression of high modernist faith. However, the twentieth-century underground mine—man-made like the modernist city, but also a part of nature like the German forests—suggests an even more compelling and revealing example.[70]

Until the post–World War II construction of semipermanent human dwellings beneath the sea and in space, the members of no other profession succeeded so well as mining engineers in creating new environments in which humans could live and work. In this sense, these mining engineers were among the world's first environmental engineers. The subterranean mine is best understood as a hybrid system that combined natural and technological systems, rather than as a completely artificial construct imposed on a natural world. Engineers worked with natural environmental systems just as much as they tried to dominate and overcome them.

Success required learning about the nature of subterrestrial water systems and airflows just as much as it required learning about the operation of Knowles water pumps and Draeger helmets. Above all else, mining engineers had to provide for the basic natural biological needs of human bodies doing hard labor.

Recognizing that mining engineers were successful subterrestrial environmental engineers also helps us to understand why many believed they could also engineer terrestrial environments. When confronted with the problem of terrestrial smoke pollution from smelters that processed the ore from their subterrestrial mines, the engineers faced the challenge with the confidence that they had the tools to solve the problem. Surely it was not too much to believe that there were terrestrial analogues to the giant Knowles pump that could "drain" the atmosphere of pollutants, or ventilation systems that could sweep smoke from the valley? Contrary to the views of some, these mining engineers and entrepreneurs were not necessarily ruthless and reckless natural despoilers who cared for nothing but profits and power. To be sure, some were concerned mostly with their own bank accounts and careers, and others had few qualms about destroying a relatively small area of the planet in the name of what they believed was material progress. However, many of these men also deeply admired the natural world and wilderness, and there can be little doubt that they made sincere efforts to minimize the pollution and environmental degradation of mining.

For these engineers, their failures often stemmed not from a careless disregard for environmental problems but rather from a dangerous overconfidence in their abilities to fix these problems. In an era when an uncritical faith in technological progress had yet to be seriously challenged, they believed that the modern ideas and methods that had served them in building mining complexes thousands of feet below the earth would also allow them to solve pollution problems on the surface. However, the mining engineers would have done well to heed the lessons of subterrestrial disasters like the Speculator fire, or of the thousands of other failings, big and small, that constantly plagued their elaborate mining systems. For all their accomplishments, their control of the underground world they had created fell well short of perfection, most tragically when it came to the systems that kept the human miners safe and alive. Their attempts to engineer the far more complex biological and ecological systems on the surface would soon prove equally flawed.

ENGINEERS IN THE WILDERNESS

When Rossiter Raymond, one of the nation's most prominent early mining engineers, first visited Colorado, he thought it "the most beautiful territory" he had ever seen, praising its "great grassy plains, with their herds of peaceful cattle, the grand mountains, the clear-flowing streams."[71] Some years later, in 1887, while returning to Salt Lake City after a hard day's work examining a mine, Raymond was struck by the sight of the Wasatch Range rising from a sea of snow: "On our side [were] orange clouds where the sun had set, the sky a delicate apple-green, an exquisite rose-tinted Abendroth on the upper half of the Wasatch Peaks, and over them in the green sky a silver, strictly silver, moon!"[72]

Reading Raymond's words today, one might assume they were penned by some late nineteenth-century nature writer. But despite his purple rhetoric, Rossiter Raymond was anything but a simple nature worshipper. Raymond, like most other mining engineers of the late nineteenth and early twentieth centuries, held a complex and sometimes contradictory view of the natural world. Many mining engineers shared Raymond's deep affection for nature and wilderness, which they believed were essential to both national character and the health of an overly civilized people. At the same time, though, engineers believed that their own identity as professionals and as men emerged from their aggressive conquest and mastery of nature. Thus, the mining engineers faced a dilemma: the more they succeeded in taming nature, the more they denied themselves the rugged wild places essential to the creation of their personal and professional ideals.

In the modern mind, the stereotypical engineer is obsessively rational and abstract, uncomfortable with emotions and concepts like aesthetic natural beauty that resist precise measurement. Spock, the Vulcan chief scientist in the 1960s American television show *Star Trek*, was an intriguing popular expression of this modern concept of the engineer. Though he was half human, Spock generally seemed incapable of emotion. Instead, he relied entirely on logic to understand the world, and his hyper-rationalism typically served the starship *Enterprise* well. Indeed, to the modern mind a spaceship itself seemed the ideal environment for this ideal engineer. The *Enterprise* appeared to be a wholly artificial environment, a world entirely cut off from nature and dependent on technological rather than natural life support systems. The program suggested the perfect environment for the engineer was one in which humans had essentially eliminated nature and replaced it with technology. The ecosphere had become the techno-sphere.

The mining engineers of the second half of the nineteenth century would have found such concepts bizarre. To them, engineering was still a humanistic profession, and the idea that an engineer should be devoid of emotion or aesthetic passions would have been foreign. Nor would they have been nearly so likely to believe—or desire—that the technological could or should be separated from the natural. Rather, the two concepts still remained closely entwined in their imaginations. As the prominent mining engineer T. A. Rickard once wrote in describing the work of the mining engineer, "The operations of nature, mechanical and chemical, are supplemented by those of man, who is a great mimic."[73] The engineers were far from alone in this. Scholars have pointed out that even the early science of ecology, which was developing at roughly this time, focused on understanding natural environments in order to improve human exploitation of them. The equation of ecology with an environmentalist ethic of preservation only began much later.[74]

For Raymond and many other mining engineers, nature was simultaneously a place to be admired for its natural beauty and the ideal venue for developing and demonstrating their professional ideals of physical strength, courage, and masculinity. This belief that the professional mining engineer should be the epitome of nineteenth-century masculinity is most readily evident in the training students received in the many new western mining schools founded during the late nineteenth and early twentieth centuries. At the Colorado School of Mines in Golden, for example, one of the first bits of school trivia memorized by the freshman mining engineer each September—just in time for football season—was the boisterous school fight song:

> Now here we have the mining man
> In either hand a gun;
> He's not afraid of anything,
> He's never known to run;
> He dearly loves his whisky,
> He dearly loves his beer;
> He's a shooting, fightin',
> Dynamitin', mining engineer.[75]

The song, of course, was supposed to be hyperbolic—it was a "fight song," after all. Yet it is also a reflection of how the student culture at Golden repeatedly drove home the idea that the mining engineer was a "true" man—virile, strong, unstoppable, and sometimes aggressive to the point of violence. Even the student dress code imposed by seniors marked the passage into engineering manhood. An article in the student newspaper warned the new student about wearing headgear inappropriate to his lowly status: "See the boy with a Stetson hat. He is not a boy, he is a Senior, and that is why he is a man, and that is why he wears a Stetson."[76]

At the Colorado School of Mines, the students equated manliness with the status of the professional mining engineer, and the chosen symbol of both was that icon of western expansion and exploitation, the Stetson cowboy hat. At times, the message went beyond symbolic expressions and clearly embraced the idea that engineers were the masculine tamers of a wild frontier. In a student newspaper editorial about a class inspection trip to some western mines, the writer celebrated the joys of hitting "the trail for the land of romance, where mines become living realities." The students, he suggests, were heading "out where the West begins, and with it, life in the great open spaces. The freedom of the West, not the movie paradise of Tom Mix, but the virile country that produced the entrepreneurs of the past decade, [and] the days when the white man took charge, exterminated the bison and broke the spirit of the Indian. The time when settlers molded camps and made a living while Mother Nature looked on and smiled." With a clear sense of nostalgia for paradise lost, the editorialist concludes, "Those were the days when men were men and romance captured the world."[77]

The Golden student culture reflected broader trends in the American culture of the day. Historians have discovered compelling evidence that many Americans were reacting against a supposed "feminization" of the nation in the decades around the turn of the century. Journalists, writers, politicians, and others claimed that Americans had become soft and effeminate during the Victorian age. Changing gender roles and the erosion of the male-dominated Victorian society of the past made many American men uncertain of their own masculinity. Some embraced various forms of what Theodore Roosevelt called the "Vigorous Life" as a remedy. Roosevelt and others urged American men to rediscover their physical and moral strength by visiting the "uncivilized" remnants of the natural world. Inspired by this "back-to-nature" movement, many men came to view the American West in particular as the ideal battlefield for recapturing their masculine identity.[78]

In the eyes of most Americans, science and technology also had a gender.

Many considered math and physics to be the "hard," masculine sciences, whereas geology, the life sciences, and other less immediately practical pursuits were "soft," or feminine. Professional engineers and geologists were new to the western states and their worth not yet fully proven in the late nineteenth century. In comparison to the mine operators and timber barons who repeatedly demonstrated their value by producing wealth and essential natural resources, the geologists and their pursuit of scientific and aesthetic interests in nature at first appeared to be economically marginal. As a result, geologists often took pains to camouflage their "feminine" affection for nature while emphasizing their "masculine" usefulness in the national drive to exploit natural resources.[79] In the same way, practically trained mine operators sometimes looked askance at college-trained engineers, with their abstract theories and emphasis on science. At the very least, the mining engineers' "feminine" appreciation for nature had to be balanced and legitimated by their useful economic role in exploiting nature.

Students at the Colorado School of Mines and other mining schools were thus immersed into a culture that tied their status as men and as professionals with the aggressive economic exploitation of a feminine natural world. Not surprisingly, such a view of nature easily translated into a deep discomfort with the idea of female mining engineers. From its inception in 1874, the Colorado School of Mines had theoretically been open to women, though it is unlikely the school administrators encouraged women to enter. In general, many engineering school administrators believed that training women was a waste of time since they were likely to marry and never put their education to use. Still, at Colorado the school would admit women if they could demonstrate their commitment to the program and were fully aware of and accepted "the conditions established in the school and on the campus." Either few women applied or few applicants met these conditions. In the eighty years from the school's founding to 1955, only four women graduated from the institution. Confounding assumptions, all four went on to pursue long and successful careers in engineering. Three of them eventually married and had children yet carried on with their careers.[80]

Four successful women graduates, though, were hardly enough to dissuade the Colorado administrators and students from the belief that theirs was a profession suited only to men. To the contrary, the students at Golden believed they were training for a special type of profession that required as much brawn as brains. They hoped this physical side of mining engineering would limit what they saw as the insidious encroachment of female students into other engineering professions. In 1927, one editorialist notes how strange women would seem within the manly culture of the mining school: "Naturally, because of our years of freedom from the gentling influences of women, Miners are rough, tough, and uncouth savages ranging the wilds of the Front Range of the Rockies." Women, he argues, would surely be uncomfortable among the hard-edged men of mines.[81] In a 1926 editorial, the author recognizes that women had begun to enter into the business world and had demonstrated their worth. Soon they would be an influence in many engineering professions, where, he admits, they had shown some aptitude for office work and laboratory research. With a nearly palpable sigh of relief, though, he concludes that because of the physical challenges of fieldwork, "mining is about the only branch of engineering that can be kept free from their influence."[82]

The school culture at Colorado thus closely linked the profession's relationship to nature and its relationship to women. In this view, the job of the mining engineer was sometimes physically taxing and required vigorous battle with an often dangerous natural world. The supposedly weaker and more timid sex was, therefore, unsuited to the profession. The mining engineer's triumph over nature highlighted his masculinity and paralleled the supposed rational and physical superiority of the male gender. This idea was far from unique to the mining engineers, and many scholars have discovered that similar ideologies pervaded the wider development of western science and technology. Mary Midgley, for example, reveals the highly gendered language of seventeenth-century scientists who spoke of mastering a female natural world.[83] Likewise, David Noble argues that modern science and technology are deeply rooted in medieval Christian monasticism. As these monastic "worlds without women" played a key role in the early development of Western science and technology, this gender division persisted, ensuring that women were widely perceived as technically and scientifically inept.[84]

Given such a historical heritage, it comes as no surprise that the emerging engineering professions of the nineteenth and early twentieth century equated masculinity and technological mastery. The tie was especially close within the more "practical" engineering professions, such as the mining engineers. As the historian Judy Wajcman argues, "All the things that are associated with manual labor and machinery—dirt, noise, danger— are suffused with masculine qualities. Machine-related skills and physical strength are fundamental measures of masculine status and self-esteem according to this model of hegemonic masculinity."[85] Likewise, Sally Hacker argues that the engineering professions emphasized scientific abstraction and technical competence and dismissed the value of human traits culturally associated with femininity such as nurturance and sensuality. The engineering professions justified their exclusion of women because women supposedly lacked both physical strength and the ability to think abstractly and rationally.[86]

The ways in which mining engineers looked at women and gender were symptomatic of the broader cultural shift toward constructing a conceptual barrier between the technological and the ecological. As technological systems became increasingly powerful and seemingly dominant, men claimed them as their unique province. Supposedly, only they had the requisite rational and analytical abilities to create and manage these complex modern technological systems. By contrast, women were increasingly associated with natural ecological systems, supposedly simple and holistic worlds suitable to their more nurturing but less incisive minds. It was, perhaps, not coincidental that Americans also began to embrace the ideal of wilderness during this time. Just as ecological woman was set apart from technological man, so too was the wilderness increasingly defined by its separation from the technological. Again, the ecosphere became distinct from the technosphere, the lower realm of the earth from the higher realm of the heavens, the wildness of nature from the rationality of the spaceship. Likewise, just as men used the power dynamics of gender to dominate women, so too did they increasingly believe their technological systems could dominate or marginalize the natural world—or perhaps even subsume nature altogether.

Historians have found evidence that this radical modern dichotomization of the human and the natural was taking place in other areas at roughly the same time. In their insightful examinations of the creation of the American national parks, Mark Spence and Louis Warren demonstrate how the new idea of "wilderness" required that humans and their artifacts be removed from places like Yellowstone and Glacier. Most tragic, even Native Americans who had long made use of these areas for spiritual and material purposes were cast out of what Americans increasingly wished to view and re-create as pristine Edens.[87] Karl Jacoby suggests a similar banishment affected many Americans who lived near parks and forest reserves and used them for subsistence hunting, wood gathering, and other productive purposes.[88] Though these authors do not cast their analysis primarily in terms of technology, it is clear that wilderness areas like the national parks were being defined in terms of the absence not so much of humans per se but rather of human technological activity for extracting natural resources.

Recent work by historians of the environment and medicine suggests a similar shift was taking place during this same period within American ideas about the human body. Conevery Valencius finds evidence that many nineteenth-century Americans believed that the health of their bodies was closely tied to the "health of the country," suggesting the continued existence of a holistic view of nature and the human place within it.[89] Likewise, Linda Nash argues that even as late as the early twentieth century, doctors in California still saw human health as a product of complex interactions between the human body and nature. At the same time, however, such holistic views that tied the human and natural together were being challenged by modern reductionist concepts of human health. As Nash explains, in this emerging new scientific view, the human body was increasingly seen as largely separated from nature. Diseases were thus caused not by a complex web of interactions between body and environment but rather by specific germs or other pathogens that managed to invade the barrier separating the human and natural worlds. A more holistic and ecological view of human health, Nash concludes, would not reappear until the post–World War II resurgence of environmental sciences sparked by Rachel Carson's analysis of pesticides in natural food chains.[90]

As professionals whose work was deeply embedded in natural and biological systems, mining engineers were struggling with precisely this same modern tendency to divide the human and the natural, the technological and ecological. Much like Nash's doctors, they initially believed that humans and technology were part of natural systems and that they could manage the combination. Nature and technology could be effectively integrated. However, as the challenges of engineering the environmental and technological interactions in the places like the Deer Lodge Valley became more complex, engineers and their corporate masters would increasingly reject such holistic concepts and embrace the modern dichotomous view that split the technological from the environmental, the human from the wilderness, and even the male from the female.

Such modernist dichotomies served a vital purpose, both facilitating and justifying the domination of a distinctly separate nature, just as they facilitated and justified the broader male domination of women in American society. What we believe is separate and distinct from us, we believe we can safely exploit, control, or marginalize. In time, the engineers would discover that these divisions were illusions as their attempts to simplify and control complex environments and technologies began to fail and ultimately collapse.

In 1911, Thomas Edison was the guest of honor at a luncheon held for the opening of an "Electrical Exposition" in New York City. According to a *New York Times* story on the gathering, many other "notables in the electrical world" attended, including Edison's longtime rival George Westinghouse. The highlight of the expo was a six-room house that had been "fitted out with everything that has been invented to make electricity do the work," from "milady's curling irons in the boudoir to a dish-washing machine in the kitchen." All the electric lights and devices were symbolically and literally linked back to Edison: the inventor had turned on the power to the expo the previous day via a remote switch wired to his lab in New Jersey.

During the luncheon, a group of "producers and consumers" of copper took advantage of the occasion to present Edison with a gift. During an earlier dinner with the inventor that year, these "copper men" had "lauded Mr. Edison, and rejoiced in the vast increase in the output due to him." Ever the good businessman, Edison had responded that it was a shame he had not had a "chunk of it." Eager to express their "appreciation of the part his inventions have played in the continuous stimulation of the copper industry," the copper men came up with what they considered the ideal symbolic gift: a solid cube of pure copper one foot on each side. Weighing 486 pounds, the copper cube was supposed to represent a "chunk" of the increased business the copper men had enjoyed over the past forty years. Inscribed on one face were the statistics for the American output of copper in 1868—the year of Edison's first patent—and for 1910. Output was just under 378 million pounds in 1868 but had grown to 1.9 billion pounds by 1910, a fivefold increase. Fittingly, one of the "copper men" present that day was Charles Kirchhoff, the president of the American Institute of Mining Engineers.[91]

After examining his heavy gift, Edison joked, "It might make a good paperweight." There is no evidence that the sixty-four-year-old inventor gave the big chunk of copper much more thought than that, though he kept the cube on display in his West Orange lab for many years to come.[92] Still, one wonders if he ever paused for a moment during his visits to the lab to study the shiny mass of copper and consider its significance as something more than a very heavy paperweight. He could not have known then about the giant stellar furnaces and supernova explosions that created and dispersed the copper he used in his electrical power systems. The scientific insights that would connect his bright little light bulbs to the lives and deaths of stars were still decades in the future. Given his lifelong interest in mining and metallurgy, he might well have understood something of the role hydrothermal deposition had played in concentrating the Earth's modest allotment of copper to a point that humans could discover and mine it. The theory of plate tectonics, however, was still unknown. A glance at the statistics engraved on the cube might have reminded him of one thing he likely already knew: that he owed as much to the leaders of the copper mining and manufacturing industries as they did to him. The fivefold increase in production made possible by Marcus Daly, Daniel Jackling, and others had provided the masses of cheap copper necessary to fulfill his vision of an electrified nation.

Perhaps it would be expecting too much to wonder if Edison also saw some of the costs paid in creating the 486 pounds of copper in his cube. Would he have even thought to see some pale reflection of Nick Bielenberg's dead cattle in its shiny red surface? Or later, after the Speculator mine disaster had been front-page news across the nation, of the 164 Butte miners killed in one of the safest and most technologically advanced underground mines in America? If Edison did consider such dark thoughts, they were likely fleeting. Like the engineers who had created those copper smelters and mines, Edison had an unshakeable faith in the power of science, technology, and invention to solve all such problems. It was only a matter of dedicating the necessary time, will, and effort.[93]

Though somewhat naïve in retrospect, such optimism was not altogether unwarranted. The electrical power system that Edison had pioneered would play a starring role in the success of one of the greatest technological fixes of the twentieth century: a machine that could actually capture smoke. But just as attempts to engineer a cleaner and safer subterrestrial environment in Butte contributed in unexpected ways to the Speculator disaster, so too would efforts to solve the smoke pollution problem in the terrestrial environment of the Deer Lodge Valley produce other unforeseen consequences. In time, the human dead zones below would find haunting echoes in the natural dead zones above.

THREE

The Stack

And they said one to another, Go to, let us make brick, and burn them
thoroughly. And they had brick for stone, and slime had they for
mortar. And they said, Go to, let us build us a city and
a tower, whose top may reach unto heaven.

—Genesis 11:3–4

The city of Anaconda is related to Butte but resembles it not at all, for it
reaches skyward with the world's tallest stack, from which the gases
of the smelter are vented 585 feet into the Montana sky.

—"Anaconda," *Fortune* (1936)

The first stop of the specially chartered train that Saturday morning was in
Garrison, a tiny farming and ranching community at the far northern end
of the Deer Lodge Valley. The train then headed south toward Anaconda,
pausing at every small country rail stop along the way to pick up passen-
gers. Everyone rode free that day, courtesy of the Anaconda, or the Amalgamated Copper Mining Company, as it was now called. When the train
reached the stop at the company's two-year-old Washoe smelter complex,
company president William Scallon welcomed the crowd of nearly a thousand farmers, ranchers, and their families. A rare photograph from that
summer day, July 25, 1903, suggests a mostly eager and attentive audience
that seems to crowd forward, curious to hear the words of Scallon and the

Washoe managers. A few, though, hold back, looking away toward the distant mountains or down to study their boots. Almost all are dressed in their Sunday best, as if they were on a church outing or a Fourth of July picnic. Most of the men wear suits and ties, the women long proper skirts and blouses. The girls in their bright summer dresses and hats stand out, floating like flecks of white in a sea of darker shades. A few women have opened their parasols, seeking some meager protection from the already hot Montana sun.

The Anaconda-Amalgamated had invited the Deer Lodge Valley farmers and ranchers to the smelter for a tour of its brand-new three-hundred-foot smokestack and flue system. In a few more weeks, the Washoe officials explained, the smoke problem that had been killing the cattle and horses in the valley during the past year would be over. Much of the arsenic dust and other pollutants would settle out in the big new flue, while the tall stack would carry whatever remained so high into the air as to render it harmless.

Over the next few hours, Washoe officials escorted their guests on a tour of the refurbished smelter. The bolder among them rode lifts up to the rim of the three-hundred-foot stack, where they could enjoy a breathtaking view of the nearby town of Anaconda and the long, narrow Deer Lodge Valley stretching off to the north. For many, though, the highlight of the tour was the luncheon, which the company held inside the giant Washoe flue. As one longtime Anaconda resident later recalled, "The setting was one of the most novel ever seen in the west." At the point where all the smaller flues converged, the new smoke chamber opened up to some sixty feet wide and forty feet deep. Here the company placed enough tables for a hundred guests per sitting and served a lunch catered by one of the best restaurants in Anaconda. Fittingly for a copper smelter, hundreds of dazzling bright electric light bulbs lit up the space, lending a carnival atmosphere to the dark cavernous brick interior of the flue.

Exactly which company official conceived of this brilliant bit of industrial theater is unknown. Perhaps it was Scallon himself, or maybe the clever Washoe manager, Frank Klepetko. Regardless, to stage a formal sit-down luncheon right in the belly of the Washoe's new smoke flue was a dramatic way of suggesting that the company had solved the pollution problem in the valley. Perhaps many of the Washoe's guests that day believed it. Surely many must have wanted to believe it, especially those who came from the neighboring company town of Anaconda, whose survival depended on the smelter's continued existence. To witness the towering smokestack, to sit with a hundred others in the echoing chamber of the mighty flue, to hear the learned engineers explain how it all worked—even the most doubtful among them must have felt some spark of hope. Perhaps the towering brick stack really could protect the valley and their crops and animals, just as the thick brick walls of the flue they sat in protected them from the hot sun that day.

In a few months it would be harvest time, and soon the cooler days of autumn would return to the valley. By then, the giant chamber where the families had enjoyed a pleasant summer's lunch would be thick with fine dusts of arsenic, lead, and other metals. As the Washoe managers and engineers had promised, the technology really did work. The piles of potentially valuable dust testified to that. Unfortunately, it did not work nearly well enough to stop the animals from dying. The struggle to control the environment of the Deer Lodge Valley had only just begun, and the search for an effective technological fix would continue. Having mastered the challenging environments of the subterrestrial world, engineers may have been confident they could use similar tools to limit smelter pollution and control terrestrial environments. However, the contradictory impulses at the heart of the engineers' professional and personal identity would also limit their ability to create a harmonious envirotechnical system. Ultimately, the conflicting impulses within the mining engineers—which reflected broader fundamental contradictions within the emerging modern American view of nature and technology—would be embedded into the very fabric of the Deer Lodge Valley itself.

FIRE AND BRIMSTONE

In its native state, most of the ore in the underground mines of Butte was essentially worthless. Copper atoms bond readily with other elements, so copper appears in a wide variety of mineral forms. A guided tour of subterranean Butte with a skilled geologist might yield sightings of the lovely blue-green copper carbonate called malachite, or the brilliant indigo copper sulfide, covellite. Butte's most common copper mineral, though, was the less exotic and attractive chalcocite, a dark gray mineral with two atoms of copper bonded to a single atom of sulfur. Often referred to as "copper glance," chalcocite with its two atoms of copper was one of the richest of the copper minerals, containing just less than 80 percent of the metal. The second most common was enargite, a mineral that balanced out its three atoms of copper and four of sulfur with the unfortunate addition of an atom of arsenic.[1]

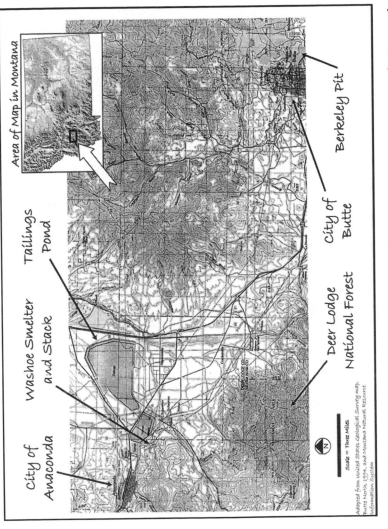

Area of Map in Montana

City of Anaconda

Washoe Smelter and Stack

Tailings Pond

Deer Lodge National Forest

City of Butte

Berkeley Pit

N

Scale = Three Miles

Adapted from United States Geological Survey map,
Butte North, 1994, and Montana Natural Resource
Information System

6. A contemporary map of the Butte and Anaconda region in southwestern Montana. For a century, the Anaconda mined copper ore from beneath Butte and from the Berkeley Pit on the surface. The company shipped the raw ore to the giant Washoe concentrator and smelter near the city of Anaconda for processing, which released immense quantities of smoke and tailings into the Deer Lodge Valley. *Adapted from U.S. Geological Survey map, Butte North, 1994.*

In these simple molecular bonds lay the root cause of the smoke pollution problem that had first plagued the people of Butte and later killed Nick Bielenberg's cattle and horses. Absent a technological means to break the copper molecules away from the sulfur and arsenic, chalcocite and enargite were not "copper ores" at all but just inconsequential pieces of rock.[2] Mere mechanical actions like crushing and sifting could never separate the copper atoms from their fellow molecular travelers. To free the valued copper required the application of intense heat—enough to split the chemical bonds and drive off the sulfur and arsenic. Chalcocite and enargite minerals were also jumbled together underground with many other common and less valuable or worthless minerals like quartz, pyrite, and granite. Engineers could partially remove some of these undesirable associates through various mechanical concentration methods. Others were removed by melting the ores with high heat. Mining engineers and metallurgists classified all of these roasting and smelting techniques under the art of "pyrometallurgy"—the making of metals with fire or heat.

Often, the first stage in making copper ore into copper metal was roasting. In the early days of mining in Butte, the companies roasted the ore by piling it with logs in giant open pits and setting it on fire. Once lit, the ore heaps (some as much as a block long) could burn for days, releasing huge clouds of arsenic oxides and sulfur dioxide into the air where, unless there were strong winds, a ring of high mountains held it in the valley. The heat and awful rotten-egg smell produced by burning brimstone—the ancient name for sulfur—apparently suggested a fitting description of the Christian hell to some biblical authors and translators. If true, then it can be fairly said that much of the time Butte once smelled like hell. The clouds of smoke were sometimes so thick that workers lost their way while walking home. The risk of collisions in the murky crowded streets was a real danger, and during severe inundations the city required carriage drivers to hire a man with a lantern to walk before them to warn pedestrians. Worse, the smoke caused or aggravated all manner of lung diseases. Death rates from pneumonia and tuberculosis were far above normal, exceeding the tolls in even dense urban areas like Chicago and London.[3]

For a time the people of Butte tolerated the terrible smog caused by the smelters, even taking a sort of perverse pride in this undeniable evidence of their city's industrial might. The early copper baron Charles A. Clark famously claimed that the arsenic dust in the air made the women of Butte among the most beautiful in the world; he apparently admired women with extremely pale skin, which can result from breathing and consuming arsenic.[4] Victorian women had sometimes used beauty products containing arsenic for that very reason. Some Butte medical authorities argued that the copper smoke also improved overall public health, acting as a prophylactic disinfectant against contagions like tuberculosis and cholera. Not surprisingly, many Butte citizens were also reluctant to attack the economic source of their own wages and prosperity. Eventually, though, the smoke pollution became so deadly that a public health movement gained enough power to demand that the smelter operators do something to solve the problem. The mining companies responded by building enclosed roasting stalls attached to smokestacks, hoping higher-atmospheric winds would sweep the smoke out of the valley, or at least dilute it enough to be harmless. The smokestacks in Butte in the 1890s were relatively modest structures: the tallest was around a hundred feet. As the mines grew deeper, the region's smokestacks would soon become much taller.

After roasting, further purification of the copper ore by smelting made the pollution problems in Butte, and later Anaconda, even worse. The verb "to smelt" comes from the German *schmelzen*, meaning simply to melt. In a copper smelter the mixture of copper and other ores is heated to a temperature of around 2,700 degrees Fahrenheit—nearly hot enough to melt iron.[5] This high temperature liquefies many of the ores and creates chemical reactions between the ore, the fuel, and the atmosphere. The copper and other valuable minerals separate out and sink below the iron, silica, and other substances, which the smelter workers can then skim or pour off. The resulting waste product is called slag, and soon tall mountains of the black granular material grew around the smelters.[6]

During smelting a host of complex chemical reactions occur, both within the liquefied ore itself and with the atmosphere, producing gases, vapors, and particles. Many of these are dangerous to human, animal, and plant life, especially when smelting the western copper ores like those at Butte. Recall that Butte's chalcocite ore has an atom of sulfur for every two of copper, while enargite has four sulfur atoms and one arsenic bonded to its three of copper. The smelting process drove some of this sulfur and arsenic out of the ore, setting them free to combine with oxygen and make sulfur dioxide, arsenic trioxide, and other noxious gases and particles.[7] In the early days, most copper smelter operations in the West simply released the arsenic and sulfur gases into the countryside as billowing clouds of stinking brownish-yellow smoke. This was the way it had been done for centuries, and Europeans had been complaining about the smoke for just as long.[8]

As the output of copper ores from the company's Butte mines grew,

Marcus Daly and his backers decided to expand his smelting operations and relocate them some twenty-six miles to the west in the Deer Lodge Valley. The company erected its first smelter in the valley in 1883 up a narrow drainage formed by Warm Springs Creek, which also provided a critical source of clean water for the operation. By the turn of the century, though, this smelter had already become too small and out of date. In 1885, the output of ore from Butte had been just under 34,000 tons. Thanks to the rapidly growing size and efficiency of the underground mines, by 1900 production had increased nearly fourfold to 127,187 tons.[9] In 1902, the company moved to meet this rapidly growing output from the mines by building the enormous new Washoe plant. Erected on a low ridge at the juncture of the Deer Lodge and Warm Springs valleys, the Washoe was just a few miles away from the old smelter. With a price tag of $9.5 million, it was one of the largest and most modern copper smelters in the world, capable of treating 8,000 (and soon 12,500) tons of copper ore every day.[10] An engineering marvel of the early century, the Washoe was an expression in iron and steel of state-of-the-art mining and metallurgical prowess incongruously set down in the remote hinterlands of Montana.

The engineers who designed the Washoe apparently gave little thought to the potential damage their marvel might do to the Deer Lodge Valley environment, perhaps because their earlier operations slightly farther up the Warm Springs drainage had not caused any serious complaints from neighboring farmers and ranchers. At first, it seemed the new smelter would be no different. The Washoe processed its first charge of copper ore on January 25, 1902, and operated through the spring and summer without problems. With the coming of fall, though, the dying and the protests began. By early November, thousands of horses and cows were dead, and the affected ranchers and farmers, Nick Bielenberg among them, were largely convinced the smelter was to blame. As one farmer noted, the new smelter "belches forth such an enormous volume of . . . poisonous gases that the mind is incapable of comprehending the vast polution [sic] of the atmosphere in its vicinity."[11]

Three years earlier, the Anaconda had become part of the Amalgamated Copper Mining Company, and it was now linked to some of the greatest fortunes and corporate holdings in American history. In 1899, two of the leaders of the giant Standard Oil trust, Henry Rogers and William Rockefeller, took control of the Anaconda as part of their scheme to establish a monopoly over the American copper industry. Their attempt to corner the copper market ultimately failed, but the Amalgamated endured until 1915, gradually establishing nearly monopolistic control over most of the mining and smelting in Butte. Marcus Daly, who had led the Anaconda since its early days as a silver mining operation and led the efforts to engineer the vast subterranean world beneath Butte, became the first president of the Amalgamated. But his influence within the complex eastern-based corporate power structure soon waned, in part because he was slowly dying from kidney disease. He was only fifty-eight when he died in New York City on November 12, 1900.[12]

The master of the Butte underground and builder of the first copper smelter in the Deer Lodge Valley did not live to see the erection of the much larger Washoe, though Daly had worked toward that goal in the final years of his life. On the other hand, Daly's premature death also meant he did not have to face the problem of cattle and horses killed by the smelter smoke. The challenges of solving the smoke pollution problem in the Deer Lodge Valley fell instead to Daly's handpicked successors. In the years immediately after Daly's death, William Scallon briefly served as president before resigning in 1904 to return to his private legal practice. For most of the first half of the twentieth century, through the years of Amalgamated control and after the company reemerged as an independent corporate entity in 1915, two men dominated the Anaconda: John D. Ryan and Cornelius F. Kelley. Even more than Daly himself, Ryan and Kelley would continue to expand and modernize the mining and smelting operations at Butte and Anaconda, greatly increasing the efficiency of extracting and processing the copper ore.[13]

It fell to Scallon to first deal with the Washoe smelter smoke pollution, though the problems would endure into the Kelley and Ryan eras as well. In the first few years, Scallon and his engineers clearly recognized and made a genuine effort to fix the problem. The Anaconda (or more precisely, the Amalgamated, though for the sake of simplicity we will continue to use the Anaconda name) first responded to the complaints by paying out more than $330,000 in compensation to the farmers and ranchers—a sum equivalent to perhaps $6 million today.[14] Indeed, the gross value of all the lands affected by the smoke was, by the owners' own estimate, about $918,000. Either way, a $330,000 damage payment in 1903 was not just a token sum. Thus, the Anaconda's first move in the "smoke wars" was a de facto (although never a de jure) acknowledgment that the smelter smoke was damaging crops and livestock.

Copper mining and metallurgical engineers had known for decades that sulfur dioxide in copper smelter smoke could damage vegetation. The first edition of E. D. Peters's influential 1887 text, *Modern Copper Smelting*, devoted several pages to the harmful effects of copper heap roasting on plants and recent scientific research on the subject.[15] Peters recalls how he had personally witnessed a passing rain cloud absorbing sulfurous smoke from a roasting operation. After the cloud had moved on another eight miles, the poisons fell "in the shape of an acrid and blighting rain upon a field of young Indian corn, withering and curling up every green leaf in the whole tract of many acres in less than an hour."[16] Exactly how the Anaconda's engineers and managers viewed the issue of arsenical poisoning of stock at this point is less clear, but the damage payments for stock deaths strongly suggest they did not deny the possibility.

The best evidence that the Anaconda genuinely tried to solve the problem was the company's decision to shut down the entire smelter for most of the summer of 1903 to install a gigantic new flue and smokestack system. At a cost of roughly three-quarters of a million dollars, the Anaconda replaced the four two-hundred-foot-tall stacks with one three-hundred-foot-tall stack located on a hill behind the smelter. The top of the new stack was now about five hundred feet higher than before, releasing the smoke nearly a thousand feet above the valley floor. The company also built a 2,300-foot-long flue with a wider cross-section designed to minimize pollution by slowing the flow of smoke to allow the heavier particles of arsenic and other poisons to settle out. This was the flue in which the valley families had enjoyed their strange luncheon on that July day in 1903. When the smelter resumed full operations in September, the Anaconda proudly announced it had responded in good faith to the smoke problem by adopting a state-of-the-art pollution control system. The problem, the Anaconda confidently claimed, was solved.[17]

The anxious farmers and ranchers of the Deer Lodge Valley watched and waited. At first it looked as if the Anaconda's flue and stack solution had worked. Clouds of sulfurous smoke reeking of rotten eggs still periodically rolled out over the valley, but for almost a year there were no further serious problems with cattle or other stock animals dying. Many no doubt breathed a guarded sigh of relief, though they may have been coughing the next second from the smoke that still remained. Indeed, the high hopes for an entirely effective technological fix soon proved unwarranted. In the fall of 1904, scarcely a year after the new flue and stack had gone online, the Deer Lodge Valley animals again began to sicken and die. This time, however, the Anaconda refused to pay any further damages, insisting its pollution control system was working and any further claims of harm were groundless. In 1905, some of the farmers and ranchers banded together to sue the Anaconda, demanding more than a million dollars in additional damage payments and an injunction against further operation of the smelter.[18] President Ryan, with Kelley as one of the company's lead attorneys, prepared to do battle in court.

The subsequent trial became one of the largest and most technically arcane suits of its kind. The farmers had two scientific experts on their side who offered compelling testimony that the smoke pollution was continuing to cause serious problems in the valley, but they proved no match for the Anaconda's impressive panel of witnesses. The Anaconda experts offered Judge William H. Hunt abundant scientific justification to doubt, if not entirely dismiss, the validity of the farmers' complaints.[19] In January 1909, Judge Hunt finally ruled in favor of the company. He based his decision primarily on the economic importance of the mining and smelter operations to the valley, state, and nation, not on the more difficult question regarding whether the smoke was actually continuing to damage the land. To the degree he attempted to sort out the tangle of competing and contradictory scientific claims, Hunt found the evidence of the farmers' experts "inconclusive" and argued the Anaconda had demonstrated it was using the "best known methods and processes" to control the problem.[20]

In subsequent years, some historians have argued that the Anaconda's experts were simply unethical and their testimony false, and they conclude that Judge Hunt's ruling was a gross miscarriage of justice. Pitting the virtuous David of the farmers against the malevolent Goliath of the Anaconda-Amalgamated, this version transforms the story into an early twentieth-century morality tale about the evils of capitalism and corporate malfeasance.[21] This interpretation is not without a measure of truth, but its insistence on rendering all in stark blacks and whites leaves it incapable of capturing the many notes of gray. The actual story is a good deal more complicated, though no less dramatic and interesting for that. The Washoe's new stack and flue system really was an example of state-of-the-art pollution control technology—at least for the year 1903. Likewise, the historian of mining and technology Frederic Quivik has discovered compelling evidence that casts the Anaconda's experts in a somewhat more positive light. The experts, Quivik argues, had been assured before beginning their investigations "that the company wanted facts" and "to know if [the smoke] was indeed causing damage to livestock in the valley." Nor were the Anaconda's experts just scientific hacks for hire, Quivik argues. Several were already among the leaders of their professions at the time of the trial, and almost all went on with distinguished careers of scientific research and publication afterward. Few of the company experts published the results of their research in peer-reviewed forums, while both of the experts for the farmers' case quickly published theirs. Nonetheless, Quivik notes that the Anaconda's experts appeared to be "absolutely convinced of the veracity of their scientific findings."[22]

On the other hand, within less than a decade it would be readily apparent even to the Anaconda that the new stack and flue had not adequately contained the smoke pollution problem. Indeed, as late as the 1960s the Anaconda smelter was releasing amounts of arsenic, heavy metals, and sulfur dioxides that, while greatly diminished from 1903 levels, were still harmful to animal and plant life.[23] Why, then, were the Anaconda engineers who built the flue and stack system, as well as the company's scientific experts who investigated it, ultimately so wrong?

Obviously, economic incentives played a major role. Like Judge Hunt himself, the Anaconda managers and engineers, and perhaps some of the outside scientists, saw the survival of the big copper mining and smelting operation as essential—to their own careers and company stockholders, to be sure, but also to local and national growth and progress. The adversarial nature of the American court system also encouraged the company to take a categorical position. An equally important and more revealing explanation, however, lay with the genuinely confident expectation that the engineers and scientists could find a technological fix for the problem. Engineers in particular embraced the idea that efficiency and new inventions would allow them to balance expanded mining and smelting operations with some measure of environmental conservation. Nowhere was this optimism greater than in mining, where engineers had already created viable, if still dangerous, subterrestrial environments. Now, the challenge facing the mining and metallurgical engineers was to extend this successful regimen of subterrestrial control further into the terrestrial world.

Evidence from other instances of smelter pollution problems occurring at about the same time as the Deer Lodge Valley case surely bolstered their technological faith and optimism. Reports of seeming success stories were beginning to accumulate during the first decade of the century. A smoke problem similar to that in the Deer Lodge Valley had plagued the copper mining district in the southern Great Smoky Mountains near the junction of Tennessee, North Carolina, and Georgia. Smelters in Ducktown, Tennessee, processed the district's high-sulfur copper ore by open-heap roasting, the same technique once used in Butte. Within a few years, the sulfur dioxide smoke transformed what had been a "beautiful, mountainous, and heavily wooded" landscape into a barren wasteland. Within a radius of several miles from the Ducktown smelters, scarcely a single blade of grass grew, and a wide swath of dead or dying broadleaf trees cut through the forest for thirty miles.[24]

Farmers and townspeople sued the Ducktown smelter operators, often winning damage payments but failing to stop the operators from releasing toxic smoke. Eventually, the U.S. Supreme Court (involved because the smoke pollution crossed the state border from Tennessee) granted the state of Georgia an injunction halting all smelting in the Ducktown area. Instead of shutting the smelters down, though, Georgia officials agreed that smelting could continue if a technological fix could be engineered to remove most of the sulfur from the smoke. European smelter engineers had already developed a successful means of capturing the sulfurous materials in copper smoke and converting it to sulfuric acid.[25] Building on this work, the Tennessee Copper Company arrived at a similar recovery system for their ores that converted the sulfur dioxide gas into sulfuric acid. The chemist Robert Swain—who would later study the smoke problems at Anaconda and in the Salt Lake Valley as well—called the Ducktown invention one of the "great industrial achievements of this country."[26] The sulfuric acid plants not only reduced the amount of sulfur dioxide released into the air, thus helping to protect the forests; they also established a profitable by-product industry for the smelter operators, who could easily sell the sulfuric acid to nearby fertilizer manufacturers. The Ducktown sulfur system was far from perfect. The smelters continued to release significant amounts of environmentally damaging sulfur dioxide and other pollutants. However, as one historian of the case rightly concludes, "the site actually represented one of industry's early successes in rectifying—given the standards of the time—a serious and legally complicated interstate air pollution problem."[27]

Smelters in Utah were also causing problems similar to those in Montana. Farmers and ranchers in the valley of the Great Salt Lake had also watched their cattle and horses die after area smelters began treating ores with high sulfur, lead, and arsenic content. In 1901, twenty-one cattle in the area died after drinking from a small creek. Analysis showed the water contained fatal levels of arsenic.[28] Eventually, some four hundred Utah farmers sued the biggest smelter operator in the valley, the American Smelting and Refining Company. In contrast to the Montana case, though, a district court judge sided with the farmers. In 1906, he issued an injunction prohibiting the smelters from processing ores with more than 10 percent sulfur content and banning the further release of any arsenical compounds into the atmosphere. Unable or unwilling to meet these limitations, two valley smelters shut down. The operators of the two remaining smelters continued to process high-sulfur ores—they had little choice, since much of the available ore in the area had far more than 10 percent sulfur. Borrowing from the Ducktown smelters, the companies instead installed sulfuric acid plants that removed much of the sulfur dioxide. To minimize the release of arsenic particles, the smelters also added "bag houses," a relatively simple (though costly) technology that captured much of the arsenic pollution in the fibers of long woolen bags.[29]

A smelter near San Francisco offered another seeming technological success story. Farmers and townspeople sued the Selby Smelting & Lead Company on San Pablo Bay over smoke problems. In 1907, a judge forbade the smelter from operating between March 15 and October 15 of every year, the period when trade winds carried the smoke inland toward coastal farms and towns.[30] Faced with the prospect of being able to operate only five months out of the year, the company installed bag houses and a sulfuric acid plant. With these pollution controls in place, the court permitted the company to resume full-time operations.[31]

In all of these cases, though, the new technologies did not fully solve the problem. The smelters continued to release sulfur, arsenic, lead, and other dangerous pollutants, though in much smaller amounts. As a result, technological optimism sometimes shaded into a sort of technological arrogance. Managers, engineers, and even supposedly unbiased government experts insisted the new technologies had adequately solved pollution problems, regardless of what some area residents might believe to the contrary. Confident in their knowledge of the complex effects of the pollutants and their technological ability to minimize it, they often considered any further public complaints to be either mistaken or duplicitous.

At Selby, for example, farmers continued to insist that the smelter was damaging their lands and their health even after the company had installed bag houses and a sulfuric acid plant. In March 1913, all parties agreed to hand the issue over to a three-member Selby Smoke Commission which promised to determine "the essential facts and lay down findings that would be accepted as final by the litigants."[32] The most prominent member of the Selby Smoke Commission was Joseph A. Holmes, the first director of the recently created U.S. Bureau of Mines. Like the many mining engineers and managers he closely worked with, Holmes believed that conservation and efficiency could simultaneously solve most pollu-tion problems, improve company profits, and ensure a steady supply of metals. Waste—of both natural and human resources—was an enemy that could be eliminated through science, good engineering, and the wise counsel of the unbiased expert. Two other scientific experts joined Holmes on the Selby Smoke Commission: E. C. Franklin, a chemist at Stanford University, and Ralph Gould, a private consulting chemical engineer based in San Francisco. Working in partnership with a variety of other special-ists in veterinary science, horticulture, medicine, and entomology, the Selby commissioners developed tests to make standardized measurements of pollutant concentrations in the air and soil. Agricultural experts con-ducted fumigation experiments, bathing farm crops in measured densities of sulfur dioxide smoke under controlled conditions of time, temperature, and humidity. Veterinarians interviewed farmers and examined the horses the men claimed had been poisoned by lead and arsenic pollution.[33]

By the time the smoke commissioners and their team of scientific and engineering experts were done, the hybrid environmental and technologi-cal ecosystem of the San Pablo Bay area had been studied in extraordinary detail. In essence, Holmes and his partners had created one of the nation's first environmental impact statements. In their attempts to understand and improve a technological system, they had also learned a great deal about a natural environmental system. The team was convinced it had discovered the full truth about how the Selby smelter smoke interacted with the farms, animals, and human bodies of San Pablo Bay. Holmes confidently states in the introduction to the commission's final five-hundred-page 1915 report that many unnecessary controversies and lawsuits could be avoided by the appointment of similar panels "of unbiased experts who, from their knowl-edge of the principles involved, will determine with precision the essential facts."[34]

What are the "essential facts" that Holmes and his partners discover once all the research has been done? Based on the evidence, the smoke commissioners find that even small amounts of sulfur dioxide could dam-age certain crop species. They also conclude that the lead from the Selby smelter had indeed poisoned at least a few horses. More important, though, the commissioners insist that the Selby engineers had fixed the smoke problem with their sulfuric acid plant, bag houses, and an additional new experimental device called an electrical precipitator. The technology was working as promised. The vast majority of the farmers' continuing com-plaints, they argue, were scientifically unfounded. Whatever problems continued to occur had other explanations. A professor of entomology at Stanford University suggests, for example, that "much of the poor condition of the trees and crops of the smoke zone is due to the work of insect pests." A professor of agronomy from the University of California implies that many of the farmers' difficulties actually stemmed from their own backward and unscientific agricultural practices.[35]

The report even offers psychological explanations for why some people continued to complain about the smoke despite the fact that the science showed the problem was solved. At least a few of the false claims were probably "actuated by ignoble motives," the commissioners suggest, as farmers tried to cash in on undeserved damage payments.[36] The majority of the mistaken complaints, though, probably occurred because the smoke-zone inhabitants had come to believe that any and all of their problems resulted from the smelter smoke. The report notes that a "sickly, anemic woman" who had been ailing for years insisted the smoke was to blame, although her doctors suggested a different diagnosis. A woman who had already lost one lung to tuberculosis told investigators she had distressing coughs and claimed the smoke made it difficult for her to breathe in the morning. Examining these individual complaints under the hard light of science, the commissioners conclude they were most likely unrelated to the smoke pollution. "So long as visible smoke is permitted to be discharged from the stacks," the report asserts, "a great many people are going to continue to believe that harm is being done—if not visible harm or damage, then some insidious form of damage that will manifest itself after awhile."[37]

Nearly a century after the commission made its investigations, it is difficult to determine just how accurate these conclusions were. However, from the vantage point of recent scientific understanding of environmental pollutants, the commissioners were probably wrong to dismiss at least some of the continuing complaints made by a "great many people." By insisting that all smoke damage claims be scientifically validated by recognized professional experts, the commission guaranteed that its damage findings would be limited by the still very modest abilities of early twentieth-century environmental science. According to the standards of more recent environmental science and public health, though, the Selby smelter clearly continued to emit unacceptably high levels of sulfur dioxide. Current toxicology references note that chronic exposure to as little as five parts per million of sulfur dioxide gas can irritate mucous membranes, cause fatigue, and produce chronic bronchitis—all symptoms that could be easily attributed to other causes. Scientists also now know that some individuals are innately hypersensitive to the gas. Likewise, modern agricultural researchers have found that very small amounts of sulfur dioxide can result in a general lack of plant health and vitality, symptoms that could be mistakenly attributed to a lack of water, temperature stress, poor soil, and a host of other causes—the very factors that the Selby commissioners identified rather than the smoke.[38]

The point of this anachronistic application of modern environmental toxicology is not so much to suggest that the Selby Smelter Smoke commissioners were completely wrong. Quite possibly some of the continuing complaints really were dishonest or mistaken. Modern evidence, though, suggests the Selby experts were just as mistaken in placing so much faith in their limited scientific understanding of a highly complex environmental and technological system. Such overweening confidence—what the geographer James Scott has referred to as "imperial pretensions"[39]—could all too easily shade into an arrogant dismissal of complaints that did not fit into their simple models.

The work of the Selby commission set an important standard for all future investigations of smelter pollution. The commission's report—published and distributed by the federal Bureau of Mines—was "widely accepted as a satisfactory method of dealing with legal controversies" over smoke damages.[40] Joseph Holmes's participation in the commission also influenced the bureau's subsequent approach to other smelter smoke controversies. Shortly before the public release of the Selby Report, the bureau published a long bulletin based on its own investigations entitled *Metallurgical Smoke*. Written by Charles Fulton, a professor of metallurgy from the Case School of Applied Science, the bulletin captured the bureau's and the mining engineers' optimistic confidence that technology was already well on its way to solving the smoke problem. "Owners of smelting plants," Fulton promises in the report, "are making every effort to devise ways and means to do away with possible damage and annoyance from smoke and are meeting with success." True, the solution was not yet fully at hand, but it appeared to be only a matter of time before it was. "As the mineral industry is one of the great basic industries of the country," he concludes, "it should be accorded freedom to work out the smoke problem to the benefit of all concerned."[41] Industry, engineers, and technology would ultimately triumph.

Given this broader historical context, it is easier to understand why the Anaconda managers, engineers, and scientific experts believed the 1903 flue and stack system was an effective technological fix. Likewise, the company managers' reluctance to acknowledge the farmers' continuing complaints can be understood, at least in part, as a failure to grasp the full complexity of the terrestrial environment they were attempting to understand and control. Given the relatively undeveloped state of the environmental sciences at the time, many of the Anaconda's engineers and experts may have been genuinely convinced that they had discovered the truth and fixed the problem. These essential failures of legibility were further compounded by the arrogance of the Anaconda engineers, experts, and managers who refused to respect or even recognize the less systematized but no less valid knowledge of the valley farmers. There was little room in their models for someone like Liz Newsome, for example, who lived about two miles from the smelter and reported that the smoke curled the feathers on her chickens and turned their combs black and yellow.[42] Or for John Bonn, a farmer who lived about twelve miles from the smelter and testified that the smoke made his cabbage, peas, and tomato plants wilt.[43]

The irony, of course, is that farmers with their highly sensitive, local, and personal knowledge of their animals and crops were pointing to a broader truth, a truth that the Anaconda experts were unwilling, as yet, to acknowledge. The effects of the chemistry shop of substances in the smelter smoke on the topographically and ecologically varied Deer Lodge Valley were far more complex and insidious than they had imagined. Even today with vastly improved tools and theories, environmental toxicology remains an exceedingly complex and difficult subject. But whether or not it could be proven in 1908 that the smoke was responsible for turning the combs of Newsome's chickens black or wilting Bonn's cabbages, the farmers' diverse evidence strongly suggested that the Anaconda's initial environmental model and solution were inadequate. To be sure, the drive for profits, personal advancement, and national progress also kept the managers from seeing this truth—but so did a less selfish (and easily condemned) overconfidence in their own ability to understand and fix the problem.

Nowhere were these confident expectations more keenly raised than with promise of the newest and most modern of the early twentieth-century smoke control devices, the electrical precipitator, the invention of a Berkeley scientist named Frederick Cottrell. For the Anaconda managers and engineers, the electrical precipitator was the greatest technological fix of all, the machine that would not only clean the Washoe smoke but also create a new source of company profits. In the end, though, at least some of the undeniable benefits of the electrical precipitator would be swept away in the flood of ore released by new methods of mass destruction mining.

THE MACHINE THAT COULD CLEAN SMOKE

Although trained as a physical chemist, the inventor of the electrical precipitator had much in common with the mining engineers he often worked closely with. Frederick Cottrell was a man who felt at home in both the natural and technological worlds, and he believed that modern scientific technology could solve many pollution problems. Cottrell was born in Oakland, California, in 1877, just two years before Thomas Edison conducted his experiments to invent a cheap and practical electric light bulb. As a boy, Cottrell was an enthusiast of the type of hobbies one might expect of a future chemist: photography, electricity, and telegraphy. In addition to his love of gadgets, though, he also had an early and enduring affection for botany and natural history. He made frequent botanical expeditions into the hills near his Oakland home, toting an odd two-foot-long japanned tin case of his own design that served as a lunch box on the trip out and as a specimen box on the way home.[44] As a precocious undergraduate at the nearby University of California (he finished high school and went to the university when he was only sixteen), Cottrell developed a lifelong passion for a practice he called "tramping" but today we would call backpacking. He made frequent trips into the Sierra Nevada, and he was especially fond of Yosemite and Kings Canyon. His friends said he was a powerful hiker, a man who took giant strides up steep mountain trails and earned the nickname "Mountain Goat." Cottrell especially delighted in moonlight climbs, and he remained an avid camper and mountaineer his entire life.[45]

Cottrell's interest in the problem of industrial smoke pollution began while he was studying for his doctorate at Leipzig University in Germany. There he worked with one of the world's leading physical chemists, Wilhelm Ostwald, a later Nobel Prize winner in chemistry.[46] The problem of atmospheric pollution was a lively subject of research in Germany during the time Cottrell studied at Leipzig. Particularly problematic was the sulfur smoke from mineral smelters. Cottrell may have been aware of the groundbreaking investigations on the smoke problem by Leipzig organic chemist Johannes Wislicenus and his colleagues at the Royal Saxon Forest Academy in nearby Tharandt.[47] Wislicenus's work preceded American research by several years, even though the typically small German smelters produced far less smoke pollution than the gigantic copper smelters of the American West.

After earning his doctorate, Cottrell returned to Berkeley to work as an instructor (soon to be assistant professor) of physical chemistry. There he became interested in the problems of the mining industry, occasionally supplementing his modest teaching salary by consulting on the chemistry of ore processing for mine operators in the neighboring Grass Valley.[48] Having gained a bit of a reputation as a chemist with a knack for solving practical industrial problems, Cottrell attracted the attention of the Du-Pont company, which had recently completed construction of a sulfuric acid plant on San Pablo Bay, not far from Selby Smelting & Lead. The Du-Pont managers wondered if Cottrell might be able to increase the profitability of the plant by devising a means to capture and remove the arsenic particles that interfered with the process.[49]

Initially, Cottrell tried to capture the arsenic with a centrifugal method he had invented. The device worked in his laboratory, but it failed miserably in the actual smoke stream of the DuPont plant. At about this time, Cottrell recalled an article he had read some years before describing the attempts of the British scientist Sir Oliver Lodge to use electrostatic charges and fields to combat smoke pollution.[50] Spurred on by the plague of England's infamous "fogs" (which were actually far less romantic-sounding smogs, combining fog with coal smoke and a host of other industrial pollutants), Lodge initially tried to use the process known as electrostatic precipitation to dissipate the thick fogs of Liverpool. By 1886, when Cottrell was still just a boy of seven, Lodge had already installed an experimental electrical precipitating device in a lead smelter stack in Wales. Lodge's device appeared to capture at least some of the pollutants, but the overall effect was too modest to be useful. The British scientist abandoned any further experiments and turned to more fundamental research in physics.[51]

Twenty years later Cottrell returned to the idea, reasoning that Lodge's theory of electrostatic precipitation was sound but that the Englishman's experiments had failed for want of an adequately powerful electrical generator. Thanks to the work of Edison, Westinghouse, and others, much more powerful electrical generators were now widely available. Cottrell began conducting experiments on electrical precipitation in his Berkeley lab in 1905.[52] He constructed a simple bench-top experimental device housed in a glass cylinder. Cottrell hooked it up to a powerful modern generator and asked one of his pipe-smoking lab assistants to blow a stream of smoke into the box. The men could not have been more pleased with the result. As the cloud of white tobacco smoke entered the charged field, it disappeared almost instantly. Cottrell immediately understood that electrical precipitation had the potential to become a major new smoke control technology.[53]

The practical and technical obstacles to converting a fragile little laboratory device into an industrial technology that would need to capture tons of pollutants every hour were daunting. Cottrell was not the first to use electrical precipitation to capture smoke pollutants—he always stressed Lodge's priority—but his scientific and engineering contributions in pursuit of a practical industrial device encompassed scores of small innovations and improvements as well as the development of a sophisticated theoretical understanding of the technology. The basic principles of Cottrell's electrical precipitator—or electrostatic precipitator, as it is called today—are simple. The machine operates by generating a powerful electrical field between an emitter plate and a collector plate, though sometimes these "plates" take the form of tubes inserted one in the other. When particulates like arsenic trioxide dust enter this highly charged field, they are bombarded by ions that transfer their positive charge to the particles. Just as a static-filled balloon will stick to a wall or "static cling" will make a freshly dried sock stick to a sheet, the positively charged dust particles are attracted to and will stick to the negatively charged collector plate. To be effective, the electrical field generated between the emitter and the collector must be strong. Cottrell's most important early contribution was simply to use a modern alternating current generator capable of producing electrical fields far stronger than those of Lodge's era.

With this promising laboratory-scale machine in hand, Cottrell joined with several of his assistants and financial backers to form the Western Precipitation Company in 1907. The company provided essential capital so that Cottrell could return to the DuPont sulfuric acid plant in San Pablo Bay and conduct small-scale experiments. The early results were promising. It appeared the precipitator could effectively remove some of the arsenic particles. Up to this point, Cottrell's new corporation had funded the experiments in the hope that the DuPont company would buy a license for the process simply because it improved profits. But before Cottrell had fully demonstrated the cost-saving virtues of his invention to the DuPont management, engineers from the nearby Selby Smelting & Lead Company approached him. They had heard rumors that Cottrell had invented a machine that could clean smoke and thus might be able to help them with their sulfuric acid problem. The Selby people, Cottrell later recalled, "insisted they were more in need of [the electrical precipitator] to avoid damage suits than Du Pont . . . and were consequently better clients for us to accept."[54]

By this time, in response to the initial lawsuits filed by farmers and townspeople, the company had already installed several air pollution devices, including a sulfuric acid plant and a bag house. Unfortunately, the bag house soon proved to be only a partial and very expensive solution. As the name suggests, the device worked by passing the smoke through about two thousand woolen bags whose fibers filtered out much of the particulate pollution. The corrosive smoke, however, also quickly destroyed the big thirty-foot-long bags, and they had to be replaced frequently and at great expense.[55] Under the mandate of the 1907 court ruling, the smelter would have to shut down between March 15 and October 15 of every year unless it continued to clean up its smoke. If they could not find a more economical "technological fix," the Selby smelter managers and engineers worried it might prove cheaper to accept the seasonal shutdown rather than continue to operate the costly bag houses. Desperate, the company offered to pay all of Cottrell's costs if he would agree to install an experimental precipitator at the smelter. Intrigued by the opportunity to solve a serious pollution problem, as well as the offer to fund his experiments, Cottrell agreed to drop the DuPont work temporarily and install his still unproven electrical precipitator at Selby.[56]

That the Selby managers were willing to pay Cottrell to experiment with a promising but not yet proven technology suggests how desperate they were to solve the smoke problem. In contrast to their counterparts at DuPont, the Selby managers did not want Cottrell's machine to improve the efficiency of the plant (though this would obviously be a pleasant bonus) but simply to avoid having to shut down. Such unyielding legal pressures were critical to the development of the precipitator. As Cottrell later wrote, "It appears doubtful whether [the precipitator] would ever have lived through the difficulties encountered in its early stages had it not appeared as a possible solution of the life and death struggle some plants were then making in the courts on fume-damage suits."[57]

A legal environmental mandate, not the promise of increased profits through increased efficiency, gave Cottrell the funds he needed. Could a little lab bench experiment be scaled up to capture the sulfur trioxide fumes and sulfuric acid mists in the five thousand cubic feet of smoke that poured out of the Selby stack every minute? That summer, Cottrell spent long and often frustrating days at the Selby plant, designing and supervising the construction of what he hoped would be a practical industrial precipitator. Cottrell's diaries from this time capture his intense efforts. He often worked fifteen-hour days at the plant, and many summer weekends found him at work at Selby rather than "tramping" in the hills. His Selby device took shape as a four-foot-square flue through which the smoke was routed. Negatively charged lead plates, four inches wide by four feet long, served as the collecting surfaces.[58] When Cottrell first passed the Selby smoke through the device and turned on the power, it worked, but only imperfectly. With further testing and a host of small improvements, though, he gradually increased performance. By the end of the summer, when Cottrell turned on his precipitator the dense brown clouds of smoke from the Selby chimney disappeared almost immediately, leaving behind only a few barely visible wisps.[59]

The Selby experiment demonstrated the industrial potential of Cottrell's invention. That July he submitted a patent application for the "Art of Separating Suspended Particles From Gaseous Bodies." Cottrell's patent fully acknowledged that scientists had long known that "suspended matter tends to be deposited" in an electric field. His innovations, Cottrell observed, offered much more: "a simple and inexpensive form of construction, free from the necessity of delicate adjustment of exposed parts, having easily replaceable parts demanding little attention, and consuming relatively little power." Cottrell included a basic design for a practical device somewhat different from the Selby machine, as it used a cylindrical wire mesh cage emitter suspended in a "precipitation chamber," which also served as the collector. The actual devices themselves could take many forms; Cottrell's hard-won operational and theoretical principles were the key.[60]

The United States granted Cottrell's patent a little more than a year after he applied. The development of a practical industrial electrical precipitator, though, was only just beginning. Despite his initial success in capturing the liquid sulfuric acid mists at Selby, Cottrell found that other applications presented new challenges. The Selby managers wondered if the device could also capture the particulate emissions from their roasting plant, which sent out a mixture of invisible sulfur dioxide gas, sulfuric acid, arsenic, and lead salts at the rate of fifty thousand cubic feet per minute. Cottrell tried, but for reasons he did not yet fully understand, this second precipitator failed.[61]

Over the next few years, Cottrell continued to refine his invention while also developing a deeper theoretical understanding of how it worked. He discovered, for example, that unlike captured liquid sulfuric acid mists that trickled down off the collecting plates, dust particles of arsenic, lead, and other metals often adhered to the collector and clogged the device. To control the problem, Cottrell devised a simple mechanism to give the plates a sharp blow every few minutes and knock the accumulated dust into hoppers beneath. Finding the ideal voltage and ionizing current for different smoke streams was also difficult. Cottrell's experiments revealed that the precipitator operated best at a voltage just below the point where sparks began to jump across the gap between emitter and collector, a phenomenon known as arcing. Hitting just that right voltage by trial and error was tricky. If he accidentally pushed the voltage too high and electricity began to arc, the precipitator could be damaged very quickly. Finding the perfect voltage and current level forced Cottrell and his assistants to develop theoretical models to predict the behavior of various particle sizes under different flow conditions, pressures, temperatures, and rates. Every installation thus demanded careful adjustment and adaptation of the precipitator to the site's unique conditions.[62]

Cottrell's greatest challenge, though, lay with the immense amounts of noxious smoke produced by the giant copper smelters of the American West. In 1911, the managers of the nation's biggest smelting trust, the Guggenheim-controlled American Smelting and Refining Company (ASARCO), came to Cottrell for help in solving smoke pollution problems at operations in Utah and California. Recall that in 1906 a district court set strict limits on sulfur dioxide and arsenic emissions from Salt Lake Valley smelters. In response, several smelters simply shut down, but the managers of ASARCO's Garfield plant, which processed the ore from Daniel Jackling's nearby Bingham Pit, attempted to keep the plant emissions beneath the mandated limits while continuing to treat high-sulfur copper ores. As the Selby engineers had discovered earlier on a smaller scale, wool bag houses worked but were expensive. The sulfuric-acid-laden smoke from the Garfield copper smelter ate through a set of woolen bags in a mere ten hours. The company's Balaklala smelter in Shasta County, California, was facing similar problems. In addition to smoke suits from area farmers and ranchers, the U.S. Forest Service charged that the smelter smoke was destroying trees in the nearby Shasta National Forest.[63]

Throughout the summer of 1911, Cottrell studied the Balaklala problem while other engineers from his Western Precipitation Company directed experimental work in Utah.[64] During these two installations, Cottrell and his assistants worked out most of the fundamental difficulties faced in using the process in the big western copper smelters. Experiments revealed, for example, that sulfur trioxide particles in the Garfield and Balaklala smoke formed a nonconductive coating on the collector plates that eventually brought the precipitation process to a halt. To combat the problem, they injected small amounts of water into the smoke stream that maintained conductivity. Cottrell and the smelter operators also learned to collect different types and sizes of particles at specific zones in the flues and stacks where temperature and other conditions allowed the precipitator to operate most efficiently. The Garfield installation resulted in a more efficient design for the precipitator itself that used five-inch open-ended pipes with wires running down their centers as collectors rather than flat plates.[65]

After a summer of intensive investigation, experimentation, and innovation, Cottrell's company installed seven large precipitators at the Garfield plant. Each precipitator contained 360 five-inch by ten-foot pipes, and combined they could treat 1.5 million cubic feet of gas per minute. When the Garfield operators switched on the power for this immense bank of precipitators, it worked almost flawlessly. Along with a sulfuric acid plant and some other pollution controls, the precipitator allowed the Garfield smelter to continue operations. Cottrell's success at Garfield no doubt came as a great relief to Daniel Jackling, since he depended on the smelter to process the ore from his mass destruction open-pit mine in Bingham Canyon. Had the smelter been forced to shut down, the survival of Jackling's mine would have also been threatened. Cottrell's installation at Balaklala was equally successful, at least in technical terms. However, since the precipitator could capture sulfuric acid mist but not the sulfur dioxide gas, it did not completely solve the Balaklala pollution problem. In any event, the district was already in decline, as the modest copper deposits were nearing exhaustion. During a period of low copper prices, the smelter later shut down.[66]

The 1907 Selby precipitator had cleaned five thousand cubic feet of smelter smoke per minute. The enormous 1911 Garfield system cleaned three hundred times that amount: 1.5 million cubic feet of gas per minute. If there had been any lingering doubts, the Garfield installation demonstrated that Cottrell's precipitator had the potential to greatly minimize (though not entirely eliminate) the pollution problems created by the gigantic western copper smelters. The vast amounts of metallic dusts and sulfuric acid liquids collected from the smoke also offered a new source of income for the smelter operators. More than any other technological fix to the smoke pollution problem, Cottrell's precipitator seemed to fulfill the mining engineers' confident expectation that efficiency and conservation could save nature and improve profits in the mining industry. In some cases, however, as will be evident in the final chapters of this book, the actual results were far less happy.

While his invention came to be seen as a smart means of increasing profitability rather than just a pollution control device, Cottrell became highly critical of the profit-oriented approach to the smoke control problem. In the early years, Cottrell had sought a healthy return from his precipitator business to supplement his modest university salary and pay off some sizeable debts. Cottrell also had a deeply idealistic desire to serve society, however, and he dreamed of somehow making scientific research pay for itself.[67] Like many engineers and scientists of the time, Cottrell was intrigued by the Progressive technocratic ideology that regarded scientific and engineering experts as the ideal leaders for the modern industrial world. In a 1937 lecture, "The Social Responsibility of the Engineer," Cottrell commended the work of Thorstein Veblen to his audience of engineers, especially Veblen's book *The Engineers and the Price System*. Cottrell argued that Veblen was exactly right: engineers were becoming mere cogs in a corporate machine directed by businessmen whose goals were profits rather than public service. Technical experts must escape this corporate domination, Cottrell believed, and use their unique skills to push corporations to become more socially responsible. Unlike some of his colleagues, Cottrell did not argue for a true "technocracy," a government controlled by engineers and other technical experts. Instead, he encouraged the engineering professions to develop the independence necessary to guide the giant corporations toward achieving the "finer and more intangible" goals of human fulfillment. No revolt of the engineers would be necessary.[68] Rather, each individual engineer working in his corporate niche should "quietly and tactfully" resist the undue influence of the profit motive—even at the risk of career advancement. Corporate loyalty need not be abandoned, Cottrell argued. But it must be balanced by increased allegiance to the professional engineering societies with their ideal of disinterested service to the public.[69]

The engineers who heard the speech that day in 1937 knew these were not just empty words for Cottrell. The inventor of the electrical precipitator had begun trying to realize these ideals nearly a quarter century before. In February 1912, just a few months after the brilliant success at Garfield demonstrated his improved precipitator might be one of the most lucrative inventions of the twentieth century, Cottrell did something truly extraordinary. He donated all of his patents for the precipitator and its many improvements to a nonprofit corporation he created with the aid of the Smithsonian Institution. Cottrell hoped this new Research Corporation, as he called it, would offer an example of how independent engineer-led businesses could place public service above the narrow pursuit of profit.[70]

Cottrell even promised that the leaders of the Research Corporation were fully willing "to risk or even sacrifice profits from the licensing or operations of patents or developments if thereby a more important public service can be rendered."[71]

In the years to come, the Research Corporation achieved many of Cottrell's goals. The nonprofit corporation not only used the precipitator to maximize the public good of cleaner air but also donated the substantial profits to support scientific research in many other areas. Somewhat akin to open-source code programming today, the Research Corporation sparked rapid improvements in electrostatic precipitation technology by allowing any interested party to experiment with the device at little or no initial charge. Under the direction of the Research Corporation engineers, industrial polluters of all sizes and types adapted and improved the device. To mention but one example of hundreds, a cement plant in Riverside, California, adapted the precipitator to capture thousands of tons of lime and clay dust that had previously coated the neighboring countryside with an impermeable layer of white dust.[72] The Research Corporation and its partners eventually adapted the precipitator to capture pollutants from a wide spectrum of industries. By 1970, there were an estimated 4,100 precipitators operating in the United States and thousands more around the world.[73]

That Cottrell's greatest early breakthroughs with the precipitator came in the mining and smelting industry was no coincidence. By the early twentieth century, air pollution from copper and other types of mineral smelters presented one of the greatest environmental threats in the nation—especially in the big western mining districts. Cottrell, of course, was not a mining or metallurgical engineer. He had not reveled in the heady achievements of creating underground environments. However, he clearly shared the mining engineers' deeply held faith that technology, efficiency, and conservation could control terrestrial environments and solve the pollution problems created by mining. In his first 1908 patent for the electrostatic precipitation process, Cottrell had even suggested his invention could be put to use "destroying fog and mists in the open air, both on land and water."[74] A man who contemplated using technology to rid seaside towns of fog and mist obviously did not suffer many serious doubts about human abilities to control the natural environment. Cottrell was also highly pragmatic, though, and he understood that any successful method of pollution control could not be excessively expensive to industry. Cottrell admitted, for example, that if every western sulfide ore smelter used the Ducktown method and began manufacturing sulfuric acid, the resulting glut of acid would overwhelm the existing market.[75]

Unlike most mining engineers and managers, however, Cottrell tempered his belief in the human ability to use technology and efficiency to control the environment with a deep skepticism about the profit-seeking drive of capitalism. Absent a strong corporate commitment to the goal of public service, Cottrell feared corporations would often use conservation and efficiency simply to increase profits, not to serve the greater good. In the years to come, though, many conveniently forgot Cottrell's warnings, insisting the electrical precipitator proved that corporations and America could have it all. Technology would provide increased profits and greater American affluence while also minimizing or even eliminating any environmental consequences.

Cottrell was right to have been worried.

THE BIGGEST SMOKESTACK IN THE WORLD

The trajectories of Cottrell's new electrical precipitator and the Deer Lodge Valley smoke problems first converged in 1907. That year, Anaconda's chief chemist, Frederick Laist, contacted Cottrell for advice. Cottrell already knew Laist—the two had been classmates as undergraduates at Berkeley in the early 1890s. Laist explained that the Anaconda was in the midst of a legal battle with area farmers and ranchers over the Washoe smelter pollution. He had heard about Cottrell's experiments with electrostatic precipitation at the DuPont and Selby plants. Laist wondered if the technique might be capable of capturing arsenic dust and other particulate pollutants from the Washoe. Cottrell had not yet patented his invention, but in his characteristically generous manner he gave Laist permission to experiment with an electrical precipitator at the smelter. In exchange, Cottrell only asked that his old classmate "keep an eye out at Anaconda" for possible profitable uses of his invention.[76]

Laist did carry out some small-scale experiments with the precipitator, but for reasons he was unable to understand, the technique largely failed.[77] In any event, within a few years the pressures on the company to find a solution to the smoke problem had diminished. Company managers felt confident they were winning the farmers' pending smoke pollution suit in the district court. The smelter manager, E. P. Mathewson, wrote to Cottrell in early 1908, more than a year before the final court ruling, "This relieves us of the necessity of doing anything further in the way of smoke damage prevention." As an insurance policy against both the pending and future litigation, Mathewson did continue to experiment with electrical precipitation, though at a slower pace. But he warned Cottrell not to expect a hefty contract in the near future. The Amalgamated executives who controlled the Anaconda, Mathewson wrote, "live a long way from here and they have no idea of the value of your apparatus." If the precipitator did not quickly provide improved profits, the eastern executives would just as quickly withdraw their support for the experiments. Mathewson explained, "They feel confident they have won the smoke suit and there is nothing further to be gained by experimenting along the lines of your invention."[78]

Cottrell wrote back to point out that the precipitator's ability to capture arsenic particles alone could result in an additional income of thirty thousand dollars a month if the poisonous material was sold to the insecticide industry. The machine would essentially pay for itself.[79] Yet, despite Cottrell's efforts and the sympathetic interest of Laist and Mathewson, serious attempts to use electrical precipitation at the Washoe languished for several years. What small-scale tests the company did make all focused on recovering more profitable materials, like copper and sulfuric acid mist, rather than the poisonous arsenic dust.[80]

Meanwhile, as the Anaconda and Amalgamated executives had predicted, the Deer Lodge Valley farmers' legal efforts were slowly collapsing. Following Judge Hunt's 1909 ruling, they turned to the U.S. Court of Appeals for the Ninth Circuit, but it upheld Judge Hunt's decision. The farmers' last legal hope was the U.S. Supreme Court. Unfortunately, the high court rejected the case on a mere technicality: the farmers had failed to print the required record of all the previous testimony and evidence.[81] Amounting to more than twenty-five thousand pages, the trial record was the largest ever produced in a suit of its kind. The financially exhausted farmers simply could not afford to pay the printing costs.[82]

By this point, many farmers and ranchers had given up and sold out. All was not yet lost, however. At roughly the same time the farmers' case was dying, the threat of a new federal case against the Washoe was growing. Theodore Roosevelt, who became president just a few months before the Washoe began operations in early 1902, had followed the Deer Lodge Valley smoke pollution case with interest. Evidence was beginning to accumulate that smoke from the big western copper smelters was damaging trees in Roosevelt's much beloved national forest preserves in other areas. Roosevelt instructed the U.S. Department of Justice to look into the possibility that the Washoe smelter smoke was having a similarly deleterious effect on the nearby Deer Lodge National Forest.[83]

Federal researchers quickly began to discover compelling evidence that, despite the Anaconda's new flue and stack system, the Washoe smoke was indeed continuing to harm the Deer Lodge Valley environment. In 1908, the U.S. Department of Agriculture published a report demonstrating that the smoke was killing or damaging national forest trees within a twenty-two-mile radius around the smelter.[84] That same year, a report by J. K. Haywood, a chemist in the federal Bureau of Chemistry, found that the concentration of arsenic in forage as far as ten miles from the smelter could kill cattle.[85] Early on, Forest Service scientists had even consulted with the great German professor Johannes Wislicenus, whose groundbreaking research in the 1890s may have sparked a young Frederick Cottrell's interest in smoke pollution. Since the distinguished old professor died in December of 1902, it was one of his last opportunities to further his goal of protecting forests from industrial pollutants.[86]

By 1910, the evidence accumulated by the government scientists strongly suggested that the Washoe smelter was continuing to damage the plants and animals of the Deer Lodge Valley. Just as the farmers' experts had earlier argued during the trial, the new flue and three-hundred-foot stack installed back in 1903 had not adequately solved the problem. Roosevelt had begun the federal investigations into the effect of the Washoe smoke on the national forest, but it fell to his successor, William Howard Taft, to act on the findings. In 1911, Taft's Department of Justice threatened the Anaconda with a federal suit if the company did not take steps to eliminate the smoke damage to national forest trees. Eager to avoid another long and costly trial (and this time one the company would have very likely lost), the Anaconda instead formally agreed to "at all times use its best efforts to prevent, minimize and ultimately to completely eliminate" harmful smoke and fume emissions—particularly sulfur dioxide, which did most of the damage to trees.[87]

To guide the Anaconda's adoption of smoke pollution controls, the two parties agreed to abide by the rulings of a three-member smoke commission similar to the one created for the Selby smoke issue several years later. The commissioners were to oversee further research into the problem and recommend possible solutions. If the Anaconda refused to cooperate or the smoke damages continued for whatever reason, the government would proceed to sue. As with the later Selby commission,[88] Joseph Holmes, the director of the Bureau of Mines, represented the government's interests.

John Hays Hammond, a famous mining engineer, served as the independent technical expert. Another prominent mining engineer, Louis Ricketts, represented the Anaconda.

Given this lineup, the Anaconda Smoke Commission was hardly hostile to the interests of big mining. Likewise, all three men reflected the mining engineering community's belief that technology and efficiency could solve the copper smelter smoke problems. John Hays Hammond was arguably the most famous member of that elite first generation of American mining engineers who had trained at the Freiberg Bergakademie. He had made a fortune managing Cecil Rhodes's gold and diamond mines in South Africa, where he had pioneered mining techniques for very deep levels. Hammond epitomized the heroic image of the mining engineer as a hard-driving, two-fisted conqueror of both men and nature. At the time of his appointment as a smoke commissioner, Hammond was sixty-six years old and at the apex of his career. He had been working for several years as a highly paid independent mining consultant to some of the largest mining interests in the world; the Guggenheim Exploration Company kept him on retainer for the princely sum of a quarter of a million dollars a year.[89]

Louis Ricketts was not nearly so famous as the indomitable Hammond, but he was very well known and respected among mining engineers. Ricketts was particularly famous for his pioneering work in large-scale low-grade copper mining operations in Arizona that shared some traits with the mass destruction mining technology Daniel Jackling was perfecting in Utah. He prospered through his association with the Arizona copper giant Phelps Dodge, where he was the protégé of the corporation's director, James Douglas. By the time he joined the Anaconda Smoke Commission, Ricketts had left Phelps Dodge to work for Anaconda, and he was currently managing one of the company's big mining and smelting operations in Mexico.[90]

Finally, as the director of the U.S. Bureau of Mines, Joseph Holmes represented the interests of the government and, by extension, those of the farmers and ranchers of the Deer Lodge Valley. Holmes had graduated from Cornell University with a degree in agriculture, but most of his coursework was in geology, chemistry, metallurgy, and physics. He later said he considered himself a trained mining engineer.[91] Holmes worked for more than twenty years as a professor of geology and natural history at the University of North Carolina. In 1904, President Roosevelt picked Holmes to head the new U.S. Geological Survey fuel-testing laboratory in Pittsburgh, where he investigated the energy content of coal. Inspired by

Roosevelt's campaign against waste in the mining industry, Holmes also became a prominent advocate of conservation and efficiency.[92] He was a friend and admirer of Gifford Pinchot, the nation's leading conservationist, and he shared Pinchot's belief that government experts should work in partnership with private industry to develop the science and technology necessary to exploit the nation's natural resources efficiently.[93]

In sum, all three of the Anaconda smoke commissioners were strong supporters of a rationalized and efficient American mining industry. Obviously, Hammond and Ricketts had personal financial incentives for ensuring the smoke problems did not significantly decrease the profitability of the big western copper mining and smelting operations. Likewise, Roosevelt had selected Holmes as the head of the new Bureau of Mines in part because the mining industry believed he had "sound ideas" and would work to further their economic interests, not limit them.[94] Given these close corporate ties, some historians have portrayed the three smoke commissioners as corporate shills who cared little about actually solving the Anaconda smoke problem. Such a charge, however, tends to obscure at least two deeper and more important insights. First, the commissioners genuinely believed that technology and efficiency could solve the Anaconda smoke pollution problem to the benefit of the company and the national forest. Second, Holmes, Hammond, and Ricketts all believed that the copper the Anaconda mined and smelted was essential to the growth and progress of the nation. To abandon the vast supplies of Butte copper was not an option, and the metal must be smelted somewhere. In their view, new technology and improved efficiency were the only viable means to both preserve the vital flow of copper and clean up the air of the Deer Lodge Valley.

To be sure, sometimes this faith that good mining practices could result in diminished environmental harm bordered on the naïve. The Anaconda smoke commissioners at least recognized the need for corporate guidance. Yet in 1910 the influential *Engineering & Mining Journal* had argued there was little need for a smoke commission at the Anaconda, as economic incentives alone would soon push the company to solve the smoke problems on its own. "Economy and efficiency are the watchwords of this business," the journal insisted, "and we are quite sure if anybody can show the company how it can save 1 cent per ton of ore treated, even if a large outlay of capital be involved, the company will adopt the idea."[95]

Of the three commissioners, Joseph Holmes was the least confident that the big copper companies would adopt pollution controls solely for reasons of increased profit and efficiency. Holmes knew from experience that many companies, no matter how progressive and enlightened, often needed prodding. The job of the Bureau of Mines was to educate the mining industry about the economic benefits of efficient new technologies and to offer expert assistance where needed. Under Holmes's direction, the bureau had already begun researching the smoke problem. Nearly sixty years before the founding of a federal Environmental Protection Agency, Holmes created an Office of Air Pollution within the bureau. The office coordinated smoke research projects at several new bureau experiment stations in the West to answer two basic questions: how to prevent damage to crops by smelter smoke, and how to recover and make a profit from materials being lost up the smokestacks.[96]

Holmes thought he had just the right man to lead these research efforts: his recently recruited chief of physical chemistry, Frederick Cottrell.[97] Cottrell and Holmes first met at a San Francisco meeting of the American Institute of Mining Engineers in 1910. In the words of Cottrell's biographer, the two men "immediately established between them a deep bond of regard and respect—perhaps the most harmonious contact of mind and personality that Cottrell was ever to experience."[98] Given their similar backgrounds and beliefs, they were natural partners. Both men were highly educated former academics who believed they should use their knowledge in service to the greater public good. Both had a deep faith in the possibilities of science, technology, and efficiency to solve problems and better human life. Both believed the mining industry could be a partner rather than a foe in achieving these goals.

With his doctorate in physical chemistry, background in mining technology, and practical experience with the electrical precipitator, Cottrell seemed the perfect man to direct Holmes's new research laboratory in Berkeley, California. Cottrell accepted Holmes's offer and resigned from the University of California in 1911 to take up full-time work for the bureau.[99] Under his leadership, the Berkeley laboratory rapidly became a center of sophisticated research on smelting pollution. Among Cottrell's first challenges was to develop new metrics for air pollution. The smelting industry desperately needed accurate and consistent techniques for measuring sulfur dioxide, arsenic, and other pollutants in flue gases and the atmosphere. Just as mining engineers had developed measuring and mapping techniques to "see" and control the subterrestrial world, now they needed new techniques for seeing and controlling the terrestrial environment. Indeed, the absence of reliable, standardized measurements of smelter pollutants may well have played a role back in 1903 when the Anaconda engineers claimed the new Washoe flue and stack had solved the Deer Lodge Valley smoke pollution problems.[100]

Meanwhile, Holmes set his other new western experiment station at Salt Lake City to work finding a profitable use for the sulfuric acid created by many western copper smelters. Once the technology was in place to capture sulfuric acid from smelter smoke, Holmes hoped, companies could make fertilizer they could profitably sell to area farmers.[101] This was just the sort of technological alchemy that Holmes and his engineers loved. Would it not be wonderful if the smoke that had once destroyed a farmer's crops instead ended up fertilizing those crops? Cottrell also worked on the sulfuric acid project at the Berkeley lab, partnering with a talented young Bureau of Mines metallurgist named Arthur E. Wells. Cottrell and Wells conducted a series of large-scale experiments designed to approximate practical and commercial conditions for manufacturing fertilizers.[102]

By 1911, the name Frederick Cottrell was already famous in engineering circles for one reason: the electrical precipitator. Four years earlier, Anaconda officials had told Cottrell that they did not need his new experimental electrical precipitator because the farmers' suit appeared likely to fail.[103] Thanks to the Anaconda Smoke Commission, the situation had abruptly changed. Holmes picked Cottrell to head up the research at Anaconda's Washoe smelter; he would be assisted by Arthur Wells, his valued associate from the Berkeley lab. Cottrell's job was to accurately measure the extent and nature of the pollutants issuing from the Washoe stack and determine how to stop their escape.

Cottrell now finally had a chance to prove what his electrical precipitator could do at the Washoe smelter. Even better, the Anaconda would be paying for it. Per the terms of the agreement with the federal government, the Anaconda provided Cottrell and Wells a yearly research budget of ten thousand dollars—today, roughly a quarter of a million. Cottrell used the money to set up a small research station right within the heart of the Washoe smelter. He used the new testing technologies he had pioneered earlier at the Berkeley lab, carefully measuring the Washoe smoke stream at various points in the flue and at the top of the stack. Not surprisingly, he soon discovered that the Anaconda engineers' 1903 attempt to solve the smoke pollution problem with the big new dust chamber and stack had been only partially effective. As the smoke stream entered the wide flue, it slowed, allowing a large amount of potentially harmful metallic dusts and chemical mists to settle out. However, his tests demonstrated that the particles of arsenic trioxide that had poisoned the farmers' cattle and horses were much too small and light to settle even in the most sluggish of smoke streams.[104]

A bag house could have captured more of the arsenic dust. But as had been the case at the Garfield and many other western copper smelters, the highly acid Washoe smoke would rapidly eat through the expensive wool bags. Cottrell's precipitator could potentially be much more economical. Over the course of several months, Cottrell and Wells made detailed measurements of the pollutants at every stage of the roasting and smelting process, noting where gases condensed into liquids under different temperature and humidity conditions.[105] By November 1913, Cottrell was ready to begin installing a small experimental precipitator. As he had with the successful 1911 installation at the Garfield smelter in Utah, Cottrell used the experimental results to design a precipitating system custom-fitted to the Washoe smoke conditions. Gradually, he determined the best material for the electrodes, the optimum gas speeds for treatment, and the ideal electrical current flows.[106]

By early 1914, Cottrell's experimental precipitator had clearly shown its potential. Electrostatic precipitation could capture much, if not all, of the arsenic in the Washoe smelter smoke, as well as large amounts of sulfuric acid mist and many other pollutants. Encouraged by Cottrell's initial results, the Anaconda agreed to purchase the rights to use the Cottrell process from the Research Corporation. To make the Cottrell precipitator an effective solution, though, would require that the company install a huge bank of the machines. The cost and the disruption to the smelter operations would be considerable. Worse, the Anaconda managers increasingly insisted that they should not have to capture pollutants like arsenic unless they could easily sell them for a profit. Cottrell had suspected the company might take exactly this position. After one of his first meetings with the Anaconda managers in the summer of 1913, Cottrell noted in his journal that the company management was already putting too much emphasis on the issue of profitability. "All [the Anaconda managers] took the position that the commission was appointed only to pass on the commercial practicality of improved fume processes," Cottrell wrote. The managers insisted the commission could only require them to engage in "such fume collection as was commercially profitable to the company."[107] Cottrell argued that this focus on profitability was not true to the spirit of the company's legal agreement with the government, but to little avail. The Anaconda managers apparently had their own interpretation of the key phrase in the agreement to "at all times use its best efforts to prevent, minimize and ultimately to completely eliminate" harmful smoke and fume emissions.[108]

The still popular engineering faith that efficiency could minimize pollution while also improving profits began to look a bit tarnished in the Anaconda case. The situation only worsened in 1915 as the expanding war in Europe began to drive copper prices upward, making the company even more reluctant to do anything that would interfere with maximizing production and profits. Cottrell and the Anaconda's own engineers had developed effective technological solutions to the smoke problem, but war profiteering and managerial resistance slowed their adoption.[109] The squabbling over whether arsenic recovery would be adequately profitable continued for two more years. Finally, in January 1916 the smoke commissioners demanded that the Anaconda install a four-unit pilot electrical precipitator to test the process at an industrial scale. The pilot operation went online in September 1916 on the Washoe's No. 2 roaster.[110] Despite some initial small technical problems, it worked brilliantly. Cottrell's tailor-made precipitator captured about 90 percent of the total solids in the smoke, including 94 percent of the arsenic.[111]

If there had been any doubt before, Cottrell's experimental plant settled the issue. Electrostatic precipitation could stop much of the arsenic from polluting the Deer Lodge Valley. The research Cottrell and Wells had done, though, led them to a conclusion the Anaconda found rather painful. If electrical precipitators were to clean the immense volumes of smoke produced by the Washoe, the company would also need to install an even bigger smokestack. In June 1917, the smoke commission directed the company to begin work on a bank of twenty Cottrell precipitators, which would be located at the base of an immense new smokestack nearly twice the size of the three-hundred-foot one built in 1903. The total price tag: $1.6 million, a sum equivalent to perhaps $25 million today.

Only a few months before, the United States had entered the war in Europe. Copper was in heavy demand, and the maximization of production could legitimately be viewed as a patriotic national duty. The smoke commissioners were sympathetic when the Anaconda again delayed. In 1918, though, the company finally diverted scarce labor and other resources to the demanding task of building the gigantic stack. When completed on May 5, 1919, the stack was 585 feet high, making it the largest freestanding masonry structure in the world. To support its own crushing weight, the stack walls were six feet thick at the bottom, two feet at the top. The inter-

7. Building the big new Washoe smelter smokestack in 1918. The base shown here constituted less than one-sixth of the total completed height of 585 feet—thirty feet taller than the Washington Monument. When the stack was finished in May of the following year, it would become the largest freestanding masonry structure in the world. *Courtesy Montana Historical Society Photograph Archives, Helena.*

nal diameter narrowed from seventy-five feet at the base to sixty feet at the top. The interior of the stack was easily tall enough to fit the 555-foot Washington Monument, though it would have been slightly too narrow at its base to fit the square footprint of the monument, which is seventy-seven feet on the diagonal.

Both then and now, the stack's immense height tends to attract attention. It could be seen from nearly twenty miles away in the Deer Lodge Valley. From an engineering perspective, though, the interior size of the stack was just as important as its height. The engineers did not intend the new brick tower, in contrast to the previous stack, to act primarily as a dilution device. The precipitators and dust chambers were supposed to make that old-fashioned solution less critical. Instead, Cottrell and Wells had found that the taller and wider stack with its greater smoke capacity was necessary in order to produce a more powerful draft. Just as a fireplace begins to pull air in as the hot exhaust from the fire rises up the chimney, the hot Washoe waste gases would create draft as they rose up the stack. The taller and wider the stack, the greater the potential amount of draft generated. Cottrell and Wells needed that extra draft in order to make the precipitators and other smoke control devices work most efficiently. At various points, the new flue system would introduce large amounts of colder outside air into the smoke stream. When the cold air hit the hot smelter smoke, some of the gases would condense out into mists of fine liquid particles. The electrical precipitator could not capture arsenic gas, but it could capture tiny liquid particles of condensed arsenic gas. Failure to understand this principle was another reason Laist's early experiments with electrostatic precipitation had failed back in 1907.[112]

A month after the stack was finished, the Anaconda completed construction of the big twenty-unit bank of electrical precipitators. Clustered at the base of the stack, the buildings housing the precipitators looked tiny in comparison to the big brick tower, but they were actually taller than a typical ten-story office building of the day. High-voltage power wires carried electricity up from the valley below to create the ionized fields in the machines. Each bank of precipitators had hoppers below to hold the captured dust, which were in turn designed to be conveniently emptied into railcars on tracks that ran right under the units.[113]

Cottrell and Wells had done their preparatory research well. The large-scale precipitators worked just as the pilot plant had, capturing about 80 percent of the metal-laden dust that had previously gone up the smokestack. The Anaconda was soon taking railcars loaded with dust rich in copper, gold, and silver out of the precipitators. Unfortunately, the initial installation of precipitators only captured 45 percent of the potentially deadly arsenic in the smoke. To capture most of the remaining arsenic, the company would have to install additional units. Profits, however, were still taking precedence. The Anaconda managers worried that they would not be able to sell the additional arsenic captured. The smoke commissioners and the Bureau of Mines tried to assuage these fears by helpfully searching for possible markets.[114] In 1922, though, the Anaconda Smoke Commission finally tired of the company's continuing insistence that every pollution control measure also be profitable. In their progress report for that year, the commissioners acknowledged that the arsenic collected by the additional precipitators might exceed the market demand. Further, there were "no new uses promising a sufficient development" of the arsenic market in the near future. Nonetheless, the smoke commission ruled that the

8. The completed Washoe smokestack in May 1919, with the old three-hundred-foot stack still standing. Frederick Cottrell's electrical precipitators were housed in the twenty-story-tall structures at the base of the stack. The captured dust was collected in hoppers and removed via railcars for disposal or further processing. *Courtesy Montana Historical Society Photograph Archives, Helena.*

arsenic pollution had to stop, regardless of whether it was profitable. "If necessary," the report insisted, "the excess production can be stored and carried as a questionable but possible future asset."[115]

In 1923, twelve years after the creation of the Anaconda Smoke Commission and nearly two decades after the Deer Lodge Valley farmers first sued for relief from the Washoe smoke, the Anaconda added the additional Cottrell precipitators designed to capture most of the arsenic pollution.[116] Subsequently, the commissioners were able to report that the Washoe arsenic output was less than a third of its previous level. As to the third that remained—still as much as twenty-five tons of arsenic per day—the commissioners confidently asserted it was of no further "nuisance to the outside surrounding community" and deemed the problem solved.[117]

By this point, Cottrell had ended his association with the Anaconda Smoke Commission and the Bureau of Mines. After guiding the Anaconda's earlier efforts to control its smoke problem, Cottrell had moved up rapidly in the ranks of the bureau. He became its chief metallurgist in 1916, assistant director in 1919, and finally served a short six-month term as acting director in 1920; he then left the organization and pursued a variety of public service research and administrative posts.[118] Cottrell's work with his electrical precipitators—initially as a service-oriented private entrepreneur and later as a public employee—was unquestionably one of his most enduring accomplishments. At the same time the Anaconda was gradually installing more and more precipitators, Cottrell's nonprofit Research Corporation had established the device as one of the most successful pollution control technologies of the first half of the twentieth century. The uses for the device rapidly expanded beyond capturing smelter smoke into areas Cottrell could hardly have imagined. In cement making, precipitators are used to capture particles rich in potash (potassium carbonate). Many cement plants subsequently found it more profitable to sell potash for use in making fertilizer and other products and to treat the cement business as a sideline. By one estimate, by 1953 Cottrell's precipitator was saving American industry about twenty million dollars annually by capturing valuable waste materials, many of which would have previously been pollutants.[119] In more recent decades, most American coal- and oil-fired electrical generating plants have used Cottrell's device to capture heavy-metal-laden fly ash pollution. There can be little question that Cottrell's precipitators have been of immense benefit in protecting both human and environmental health.

Ultimately, the electrostatic precipitator often did live up to the early conservationist dream that pollution control could also be good for business. Cottrell's struggle with the Anaconda, however, also revealed that a narrow focus on profitability could all too easily delay the adoption of the device or minimize its effectiveness. Cottrell had been rightly suspicious of the early twentieth-century belief that corporate goals and the public good would always harmonize. Smoke abatement motivated primarily by profits failed in other unexpected ways as well. Since the pollutants captured and removed from one area were subsequently sold for use elsewhere, they frequently ended up harming humans or the environment in surprising new ways. As Cottrell's biographer notes regarding the Anaconda's problems in finding some way to dispose of the arsenic and other poisons captured by its precipitators, "Here was a splendid example of how embarrassingly indestructible matter can sometimes be."[120]

Having been forced to invest $1.6 million to capture arsenic its managers really did not want, the Anaconda pushed its sales staff to find buyers for a substance that had been poisoning horses, cows, and people in the Deer Lodge Valley. The company's first attempt to transform poisons into products focused logically enough on insecticide manufacturers. Southern cotton growers had long known that arsenic was an effective means of killing boll weevils and other pests. Soon, though, the company developed an in-house use of the arsenic as a preservative for mine timbers, an idea initially suggested as a conservation measure by the U.S. Forest Service's Forest Products Laboratory in Madison, Wisconsin. Before this, Anaconda had used creosote as a preservative, but starting in 1925—six years after the initial Cottrell electrical precipitators began capturing tons of arsenic from the smoke stream—the plant replaced the dangerously flammable creosote (recall the disastrous Speculator fire) with an arsenical compound made from the smelter dust.

Both of these new profit-making opportunities eventually fed the arsenic captured from the Washoe waste smoke back into the environment. The precise effects of the arsenic sold for use in insecticides on southern cotton plantations and other farmlands are difficult to determine. However, because arsenic is an elemental substance, natural processes never break it down; the poison continues to accumulate in soil and water, where it can harm both animal and human life.[121] Obviously, arsenic is highly toxic to humans and many other animals when ingested in high concentrations. When farmers used the pesticide properly, such dangerous high levels rarely resulted. Yet modern toxicological research now suggests that even relatively low levels of arsenic may cause cancer or other serious diseases in humans and other animals. Likewise, the use of pesticides of any type has increasingly come under fire by scientists who argue they disrupt ecological balances that naturally keep insect populations in check and can lead to pesticide resistant superinsects.[122]

The arsenic the Anaconda used as a mine timber preservative also proved "embarrassingly indestructible." For decades, much of this arsenic that would have otherwise polluted the air of the Deer Lodge Valley was effectively returned to the underground environment it had initially come from. There it posed relatively little environmental threat once the miners installed the timbers and ceased handling them. When all mining at Butte ceased in 1982, though, engineers shut down the big pumps that had removed groundwater from the mine for decades. As the water resumed its natural level, it flooded thousands of miles of underground passages, many of them lined with the arsenic-treated timbers. Though a minor factor in comparison to the vast amounts of acidic heavy-metal water coming from the mine passages themselves, the arsenic in the timbers has contributed its share to the poisoning of the Butte groundwater, helping to create a toxic soup unfit for human consumption or for irrigation.[123]

The arsenic and other metals captured by the dust chambers and precipitators also poisoned the men working at the Washoe plant. One historian suggests that many smelter workers who handled the dust from the flues and precipitators died at a young age from lung cancer and heavy-metal poisoning.[124] Further, when the Washoe smelter finally shut down in 1980, in addition to about 185 million cubic yards of toxic tailings, the Anaconda left behind more than 250,000 cubic yards of metallic dust captured from the smelter smoke. The very ability of the precipitators and dust chambers to winnow out the arsenic from the other elements in the smoke actually helped to increase the arsenic concentrations in the dust to deadly levels approaching seventy thousand micrograms per gram. Unlike the relatively clear effects of the original arsenic air pollution, the dangers of this captured arsenic were more gradual and insidious. Over the course of decades, wind and water spread small amounts of the arsenic around the property near the smelter. In the early 1980s, state health investigators belatedly discovered that the children in the nearby town of Mill Creek had dangerously high levels of arsenic in their urine. Unable to effectively clean up the area, in 1987–1988 the Environmental Protection Agency evacuated and relocated all the residents of the Mill Creek community. Bulldozers leveled the houses, stripped off the contaminated soil, and transported it to a somewhat safer storage site in the area.[125]

All these unintended consequences of the smoke commission and the Anaconda's efforts at pollution control point to an obvious if somewhat heretical question: might it have been better to allow the Anaconda to release the smelter pollutants into the air as it had done originally? The answer depends, in part, upon who it is you ask. From the perspective of the Deer Lodge Valley farmers and ranchers who had managed to stick it out until the big stack and precipitators went online in 1923, removing a good portion of the arsenic from the Washoe smoke was obviously beneficial. For the workers emptying the precipitator dust hoppers, the far more dilute atmospheric arsenic pollution might have been preferable. The company could have easily minimized their risks, however, with proper safety procedures. The use of by-products like arsenical insecticides on southern cotton plantations spread the arsenic over a wide area far removed from the Deer Lodge Valley, though the exact effects are difficult to gauge. Still, it seems reasonable to conclude that the careful use of an arsenical pesticide for the seemingly beneficial killing of cotton boll weevils was preferable to the random poisoning of thousands of Deer Lodge Valley cattle and horses. After the Washoe began capturing sulfuric acid, the company also followed the Ducktown example and manufactured fertilizer, which sold around the nation. When overapplied, fertilizers can cause algae blooms and eutrophication of lakes. But used properly, a sulfuric-acid-based fertilizer again seems much preferable to a sulfuric acid smoke that kills crops and trees.

On balance, then, the smoke commissioners were right to insist that the Anaconda capture and reuse at least some of its smoke pollution. Yet, as the previous paragraph suggests, the issue is not nearly so straightforward as it might have seemed at first glance. A thoughtful answer demands careful analysis of how the smelter by-products were subsequently transformed and used in new environments. Strikingly, these critical issues received little or no attention at the time. In the Anaconda case, scientists and engineers created new tools and models for understanding how smelter pollutants interacted with local environments. Inventors like Cottrell developed techniques for capturing these pollutants. Anaconda and the Bureau of Mines experts found new markets for the resulting products. The efforts, though, stopped there. There is little evidence that anyone considered how the new products made from old poisons might harm humans, animals, and environments elsewhere. Of course, relatively little was known at the time about the dangers of arsenical pesticides or fertilizer runoff. Nonetheless, given what they had learned about the effects of arsenic and sulfuric acid in the Deer Lodge Valley, it is remarkable how few scientists and engineers were even curious about what these substances might end up doing elsewhere. Once a machine like the Cottrell precipitator had captured the arsenic, the effects it had after leaving the gates of the Washoe seemed of no concern.

In this light, it is clear that even brilliant technological fixes like the electrical precipitator also worked to disguise the full magnitude of the environmental challenges presented by large-scale mining. Rather than face the unpleasant reality that mining and smelting produced tremendous amounts of dangerous chemicals that were very difficult to safely use or dispose of, engineers and managers used these techno-fixes to preserve—rather than abandon or fundamentally modify—their hazardous industrial systems. Consider that today most American mining companies simply avoid ore deposits containing arsenic altogether because the element is so difficult to dispose of.[126] Or recall how a number of big western copper smelters were forced to shut down before the technology for making sulfuric acid was developed. To use modern psychological jargon that seems oddly appropriate, precipitators, bag houses, and acid planets were "technological enablers" to a dangerously sick and self-destructive industrial system.

Any truly deep cure for the system, though, would be painful for everyone—most obviously for the managers and engineers who had staked their careers and fortunes on the continued survival of big mining and smelting, but also for the mass of Americans who were increasingly dependent on a ready supply of cheap copper: the worried mother searching a parenting manual under the light of an electric bulb; the new homeowner eager to have a house whose plumbing would last longer than her mortgage payments; and, soon enough, the small boy bathed in the yellow glow of a radio tuner as he listened raptly to the adventures of his latest hero. All these consumers, whether they recognized it or not, were deeply dependent on the industrial system of mining. Ironically, even as they were becoming ever more deeply dependent on natural resources, nature itself seemed increasingly remote and unimportant, other than in its guise as a pristine wilderness for occasional therapeutic visits. After all, the long chain that linked an electric dishwasher back to a mine in the mountains of Montana was already difficult to see. How much more difficult, then, to connect a dishwasher to boll weevil insecticides in Mississippi, or an automobile radiator to the fertilizer applied that morning to the garden rose bed. With hundreds and even thousands of similar broken chains of causality permeating their everyday lives, it became easy for many Americans to believe that their modern technological environment had broken free from the natural environment altogether.

Whatever their failings and missteps, and they were many, the engineers and managers of the Anaconda and the nation's other great copper mines and smelters had given Americans abundant copper. So much copper, in fact, that the metal would at times be almost ridiculously cheap. Just as cheap oil helped drive the postwar American economic boom, so too did cheap copper keep the engines of modernization, manufacturing, and consumption humming. This mass consumption of copper products, though, required an equally powerful means of extracting the raw metal ore from the earth. After Daniel Jackling pioneered his technique of mass destruction mining at the Bingham Canyon mine in Utah, immense new reserves of copper would be instantly added to the world's mineral ledger book. However, mass destruction also made it increasingly difficult to maintain the happy illusion that efficient engineering and brilliant technological fixes could give Americans nearly unlimited supplies of copper with few if any costs. Inconvenient truths that had once been partially obscured in the subterrestrial mine or in the alchemy of making poisons into products were about to become far more difficult to ignore.

FOUR

The Pit

Strange to say, in the half-century since the advent of electricity
made copper so indispensable for our growing civilization,
new copper districts have been discovered
just when the world needed them.

—Ira B. Joralemon, *Romantic Copper*

What Power is responsible for the Universe made ore deposits;
but mines are made by the genius of men.

—A. B. Parsons, *The Porphyry Coppers*

No one knows precisely when the change began or who began it. The shift
in language must have been gradual, though the immense physical transformation itself had occurred with remarkable speed. Regardless, by the
early 1930s the big mountain of copper that people had long referred to as
"the Hill" no longer seemed to warrant that name. True, the remnants of
the old hill of copper still endured along one wall. But it would have been
readily obvious to anyone who bothered to look that the dominant topographical feature was now concave, not convex. In less than a quarter of a
century, much of the mountain had not just been removed—it had been
replaced by a rapidly growing hole in the ground that everyone now called
"the Pit." In the shift between those simple familiar one-syllable words was
captured an event unlike any that had ever before occurred in all the mil-

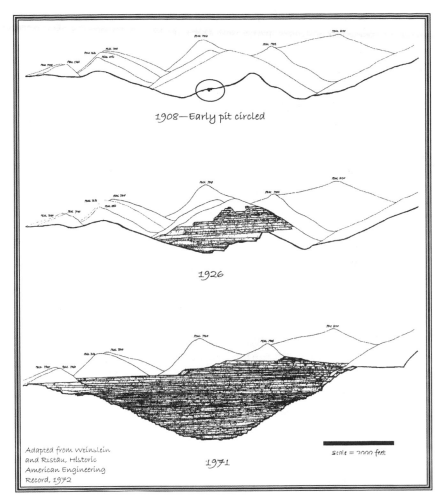

1908—Early pit circled

1926

1971

Adapted from Weinstein
and Ristau, Historic
American Engineering
Record, 1972

Scale = 2000 feet

9. A schematic profile illustrating the excavation of Jackling's Bingham Canyon mine. In 1908, miners working on "the Hill" used dynamite and steam shovels to cut into one side of the mountain of low-grade copper ore. But by the 1930s, much of the original hill of copper was gone, and miners increasingly worked in "the Pit." *Library of Congress, Prints and Photographs Division, Historic American Engineering Record, Reproduction Number HAER UTAH, 18-BINCA, 1-4.*

lions of years since human beings first began scratching in the earth with a pointed stick.

For more than twenty years, the Utah Copper Company steam shovels had been eating away at the big mountain of low-grade copper ore that lay at the head of the steep Bingham Canyon outside Salt Lake City, Utah. The creation of the Missouri-born and -trained mining engineer Daniel Jackling, the Bingham Canyon mine was the first low-grade open-pit copper mine in the world. Miners sank the first charges of dynamite into the steep flanks of the mountain in 1906. In the years that followed, Jackling's company girdled the mountain with a corkscrew of wide benches steadily tightening inward. During these early years, the miners and townspeople of the neighboring community of Bingham had referred to the mine site simply as "the Hill." By 1932, however, the managers of Utah Copper could proudly boast that they had dug out more than half a billion tons of ore and waste from the mine, and the people of Bingham could begin to speak of going to work in "the Pit." Jackling had achieved a bizarre sort of personal and human milestone. For the first time in history, human beings had done what earlier generations had thought only the gods could do: annihilate a mountain.[1]

Perhaps reluctant to acknowledge the gravity of what Jackling had done, for years Utah Copper's public relations department breezily boasted that it had done nothing more than make "a molehill out of a mountain."[2] Others, in a conscious parody of Butte's nickname as "the richest hill on earth," now ordained Bingham as "the richest hole on earth."[3] Bingham's yin to Butte's yang, as it were. But nothing captured the extraordinary reality of what had happened better than those series of unrecorded moments, repeated countless times over perhaps several years, when individual human beings who had once looked and seen a hill now looked at the very same spot and saw a pit.

In truth, Jackling and his colleagues were almost certainly the first to begin erasing the hill and seeing and talking about a pit. Years before it became an everyday topographical reality for the workers and citizens of Bingham, they had envisioned, planned, and spoken of a future pit. Such careful planning was essential if they were to avoid a tremendous waste of time, effort, and expense—and avoiding waste was at the very heart of Jackling's enterprise. Thanks to modern mapping and drilling technology, Jackling had powerful tools to help him imagine how the pit would expand in the future. He knew from the start, too, that if the mine was to be profitable, it would also have to be very big. Yet even Jackling must have been amazed at times by the titanic dimensions his pit eventually reached. When he died in 1956, exactly half a century after digging began in 1906, the Bingham Canyon pit was the largest human-created excavation on the planet, a new wonder of the modern world remarkable for what had been removed rather than for what had been created. Indeed, if that wonder of the ancient world, Cheops's Great Pyramid of Giza, were upended and dropped down into Jackling's pit today, it would barely fill the bottom sixth.

As tempting as it might be to indulge in Freudian speculations about the subconscious anxieties that drive men, past and present, to build giant structures, Cheops and Jackling did not really share the same motivations. The sheer size of Jackling's pit was never a goal in and of itself. The Bingham Pit is a monument not so much to one man's oversized (or undersized?) ego as to the bizarre consequences of pushing modern engineering technology, speed, and efficiency to their extremes. Was there some natural limit to the technological extension of scarce mineral supplies? Perhaps the supplies of necessary energy would prove inadequate at some point, or the pace of technological innovation would falter. By the late twentieth century it increasingly appeared that the limits were more likely to be environmental than technological. But in the early decades of the century, many mining engineers continued to believe that they could use new technologies and improved efficiency to limit or eliminate the hazardous environmental effects of big mining and smelting. The Cottrell precipitator, sulfuric acid capture, and seeming success stories at Ducktown, Selby, and even Anaconda offered promising signs of progress. At the Bingham Pit, though, the relentless drive for efficiency threatened to overwhelm earlier efforts to control environmental damage. Indeed, at Jackling's rapidly expanding pit the idea that the subterrestrial mine environment might be harmonized with the terrestrial environment made little sense. In the pit, the mine environment *was* the terrestrial environment, and its annihilation was inherent to the inexorable logic of mass destruction. Increasingly, it seemed almost any area of the planet could be a mine if only the energy, technology, and will could be mustered to make it so.

After all, was it not true that mines were made by men, not by nature?

JACKLING'S VISION

Today the several-square-mile area of the Bingham Canyon mine is one of the most industrialized landscapes in the United States. In addition to the yawning chasm of the pit itself, acres of huge waste piles spill down into every narrow gorge for miles around. Before the first prospectors began to scour the steep slopes, Bingham Canyon had been a cool and lushly forested wilderness, home to dense stands of red pines where the tree trunks sometimes grew three feet across. Within only a few decades after mining began in the canyon in 1863, the gold and silver miners had stripped away the magnificent pine tree forest for fuel and timbers.[4]

In the centuries before, Native Americans had occasionally visited and hunted in the Bingham Canyon, leaving little discernible trace of their brief stays. But the earliest Euro-American occupants, mostly followers of Brigham Young's religious community based in nearby Salt Lake City, came to Bingham Canyon to stay. These Latter-day Saints, or Mormons, put their considerable communal energy into the building of farms, towns, and irrigation works. The Mormon leaders were not, as some historians have argued, opposed to mining as a matter of principle.[5] Although church elders did not actively encourage mining for gold, silver, copper, or lead, the church did sponsor exploration missions for minerals they considered essential, like coal and iron, and developed promising prospects.[6]

Mormon settlers recognized the signs of copper deposits in Bingham Canyon as early as 1860, but as the *Deseret News* noted at the time, "a man who would engage in copper mining in an inland country like this, might, by some, be considered in a state of insanity."[7] During the Civil War, the federal government posted a contingent of soldiers at Salt Lake City to monitor what some northern politicians considered to be potentially rebellious Mormons. Colonel (later General) Patrick E. Connor disliked the Mormons, and he encouraged his men to prospect for minerals in the Salt Lake area in hopes that a rush of gentile gold-seekers would inundate the state and bring a quick end to Mormon domination.[8] Although a California-style gold rush never materialized in Bingham Canyon, an optimistic miner did ship the first carload of copper ore from the canyon to Baltimore for processing in 1866.[9] Serious development of the Bingham copper had to await big money, which finally arrived in 1899 when John D. Rockefeller and the other Standard Oil interests put up the capital to develop the Highland Boy mine, a rich deposit of 12 percent copper ore shot through with healthy amounts of gold and silver. The same financiers soon formed the American Smelting and Refining Company and erected three smelters in the Salt Lake Valley. In 1901, the Guggenheim family purchased a controlling interest in the company.[10]

When a young Daniel Jackling first visited Bingham Canyon in 1899, the area was already a profitable producer of copper from a handful of small high-grade deposits. But Jackling had not come to Bingham to examine the rich copper-gold mines that everyone recognized as moneymakers. He had a very different type of mine in mind: an immense mountain of ore that averaged 2 percent copper or less and contained only small amounts of gold and silver. To the copper miners of the day, this was far from a promising prospect. Many experts questioned whether it was a copper ore deposit at all. Jackling's previous mining experience, his grasp of the economics of high-speed throughput production, and an intense desire for wealth and success made him different. Where others saw only worthless rock, Jackling thought he saw a fabulous copper mine.

Daniel Jackling, born in 1869 in the tiny town of Hudson in western Missouri, had not enjoyed an easy life. His father died just four months after his birth. His mother followed two years later.[11] The orphaned boy spent the next sixteen years of his life shuffling between relatives. He earned his own keep from an early age by driving plow teams and working other farm jobs. An intensely private man in later life, Jackling rarely spoke of his boyhood life as an orphan. After he became a multimillionaire, he gave one pair of foster parents a farm and a town house. Yet judging by the handful of surviving letters Jackling wrote to the couple, his relationship with them was strained and formal, his gift more dutiful than affectionate. Warm family ties and a sense of belonging were probably rare in Jackling's early life.[12]

Jackling initially planned to become a teacher. He had no deep interest in educating the children of Missouri farmers, but he saw teaching as a practical means of earning enough cash to realize his dream of owning a farm. He enrolled in the Missouri State Normal School while continuing to work on an uncle's farm. But a chance encounter with a civil engineer using a transit to lay out a Missouri city street inspired him to abandon both teaching and farming. Jackling later recalled being utterly captivated by the precision of the surveyor's tools and craft. Given the importance of mapping and measurement in the development of large-scale mining operations, Jackling's fascination was telling. Perhaps the controlled and controlling task of mapping appealed to a young man whose life had often been so chaotic and unstable. Perhaps it just appealed to a sharp but as yet untutored analytical mind. Regardless, according to Jackling's own account his fascination with surveying inspired him to enroll in the Missouri School of Mines with hopes of becoming an engineer.[13]

What Jackling discovered when he arrived on the little campus in Rolla could hardly have been very inspiring, though. Founded in 1870, the school was near the Missouri Lead Belt, an important source of the soft metal since the eighteenth century. The town of sixteen hundred had no running water or sanitary sewer system, while the school itself had languished from poor funding and enrollment from its inception.[14] The president of the more prestigious Michigan School of Mines to the north observed, harshly though probably not unfairly, that Rolla was "unworthy of being called a mining school" and was little more than "a country academy." The school's reputation improved significantly after George Ladd, a Harvard Ph.D. in geology, became president in 1897.[15] But by then Jackling had already graduated. The truly prestigious programs at Columbia, MIT, or even Michigan were almost certainly beyond Jackling's reach in any event, and whatever its failings, Rolla did provide at least a basic education in mining engineering. Just as he had done throughout his life, Jackling earned his own keep while in college, working summers surveying for the railroad to pay tuition.[16]

Jackling graduated in 1893 with a bachelor of science in metallurgy. He remained at the school for a short time teaching chemistry and metallurgy but was ambitious for something grander and more adventurous than the life of an academic. In 1894, he quit his secure but unexciting teaching job and set out to make his fortune in the West. He was just in time to join one of the last great western gold rushes. Three years earlier, prospectors had discovered a huge deposit of complex gold ores at Cripple Creek, more than 9,400 feet high in the Rocky Mountains of central Colorado. The resulting rush was a bit of a historical anomaly, a modern restaging of the classic western gold rushes that were, by then, already four decades into the past. When the tall and lanky young Jackling (at six foot three, Jackling towered over the average five-foot-eight man of the day) arrived in Cripple Creek, it must have been as if he were stepping back in time.[17] Saloons and brothels lined the rutted dirt streets, and young men with more big dreams than knowledge and experience were thick on the ground.

Jackling did not even have enough cash for the stage trip from Divide, Colorado, up to the remote mining camp. He convinced a friendly stage passenger to take his meager luggage and made the eighteen-mile journey on foot. When he arrived in Cripple Creek he had three dollars in his pockets and no job.[18] But college-trained engineers were a rarity in the frontier mining camp, and even Jackling's degree from the impoverished little Missouri School of Mines must have seemed impressive. At the very least, it eventually helped win him a job at a gold mill operated by a Colorado mining entrepreneur and adventurer named Joseph Rafael De Lamar.[19] Joseph De Lamar was a Dutch immigrant who had been a master sailor, a ship's captain, and a skilled underwater excavator and ship salvager in his younger years. When he was thirty-five, De Lamar abandoned the risky life of an above- and underwater seaman, though he did not stray too far in a sense. He took a six-month course in mining engineering and metallurgy at the University of Chicago and set out to pursue subterranean mineral wealth in the American West. With a lifelong interest in mechanical technology and scientific research, De Lamar was the ideal boss for an ambitious young man like Jackling. Seventeen years his senior, De Lamar must have seen some spark of inventive talent in Jackling. He encouraged Jackling to make technical experiments and increasingly turned to the young metallurgist for assistance with his rapidly growing portfolio of mining projects. In 1896, only two years after the men had first met, De Lamar demonstrated his confidence in Jackling by putting him in charge of designing and building his new gold mill in Mercur, Utah. The Mercur mill was one of the first in Utah to use the revolutionary new gold cyanidation process and among the first mills anywhere to use electrically powered machinery. Jackling's design was a success, and De Lamar profited handsomely, becoming one of the wealthiest men in the nation. In 1906, he built a spectacular Beaux-Arts style mansion on Manhattan's Madison Avenue, loudly but tastefully announcing to the New York elite that the seafaring Dutch immigrant had definitely arrived.[20]

At the time of his triumph with the Mercur mill, Jackling was twenty-nine years old. Like Rossiter Raymond, John Hays Hammond, and many other famous mining engineers, he had proved his technical mettle while still a young man. Also like these famous men, Jackling had no desire to spend his entire life working for others. He wanted his own fortune, his own magnificent mansions. While completing construction on De Lamar's Mercur cyanide mill, Jackling thought he might have discovered his chance to make the leap from skilled technical wage earner to independent mining entrepreneur. In 1899, De Lamar sent Jackling and Robert C. Gemmell, a graduate of the Michigan School of Mines, to look over a two-hundred-acre mining claim in Bingham Canyon owned by a successful Utah gold miner named Enos Wall.[21] De Lamar wondered if Wall's copper deposits might have enough gold to be profitable using the new cyanidation process Jackling had installed at the Mercur mill. De Lamar also asked the two men to consider whether the low-grade copper ore might supplement the profits from the gold.[22]

Jackling and Gemmell's first task at Bingham was to estimate the size and the ore grade of the property. Everyone knew Wall's deposit was big—the surface signs of copper were obvious over many acres. But what happened to the ore beneath the ground? How far did it extend beyond the boundaries of the surface indications? There were extensive underground tunnels in the hill already, but they had been built to excavate the higher-

10. A young Daniel Jackling, on left, poses with two business associations in 1898. The following year, Jackling would visit the Bingham property for the first time. There he and his partner, Robert Gemmell, conceived an audacious plan to mine an entire mountain of low-grade copper ore using steam shovels. *Courtesy Special Collections Department, J. Willard Marriott Library, University of Utah.*

grade ore in the mine, not to explore the low-grade material everyone had previously assumed was worthless. Jackling and Gemmell needed more information. Just a few years before, making expensive human-sized tunnels would have been their only option. Fortunately, by 1899 the team could use a new underground sampling technology called a churn drill, a revolutionary device examined in more detail later in this chapter. Gemmell, who directed the sampling operations, was able to use churn drilling to sample the ore hundreds of feet below the surface at a fraction of the cost of sinking a shaft.[23] His churn drill holes were among the first to probe the Bingham ore body at depth. With the samples from the existing tunnels and Gemmell's churn drill material, Jackling then set to work on testing.

He built a small experimental mill near the site using the best machinery he could find for concentrating the extremely low-grade copper ore. Jackling's work would answer the central question: could he design a process to increase the percentage of copper enough to make the ore profitable to mine and smelt?

Based on the churn drill samples and the concentration mill tests, Jackling and Gemmell came to a bold and even seemingly bizarre conclusion. De Lamar had sent the two men to Utah to see if Bingham might make for a profitable gold mine with some supplemental income from copper. But in their September 18, 1899, report, they suggested that he should do exactly the opposite: develop the Wall property as a massive copper mine in which the small amounts of gold would just sweeten the deal. Jackling and Gemmell argued the big low-grade copper ore deposit could be the main source of profits if it were excavated using steam shovels and trains, moving very large amounts of ore very rapidly through a very big concentrating plant.[24]

De Lamar certainly trusted the judgment of the two men, and Jackling had proven his worth with the Mercur mill project. His men had also done their work carefully, and the churn drill samples and mill test results were promising. In the end, though, De Lamar balked. Even if Jackling and Gemmell were correct that the property could, in theory, be profitable as a copper mine, De Lamar thought the proposition was too risky and the need for start-up cash too great. The two young engineers estimated they would need at least three million dollars in up-front capital to build a fifteen-mile-long mountain railroad and the concentrating mill—all this before the mine produced even one cent of profit.

De Lamar, the former sea captain who had thrived on risky propositions, was not being unusually timid. Many other mining engineers and financiers ridiculed, or at least sharply questioned, the Jackling-Gemmell report. An editorial in the influential *Engineering & Mining Journal* was especially discouraging. "It would be impossible," the editors warned, "to mine and treat ores carrying two per cent. or less of copper at a profit, under the existing conditions in Utah." The editors' conclusion was brutal in its certainty: "There appears to be no doubt as to the worthlessness of the proposition."[25] In fact, the editorial was actually referring to the Boston Consolidated copper mine, a neighboring operation further up the hill, where Samuel Newhouse, a New York lawyer turned mining financier and operator, was promoting a scheme somewhat similar to Jackling and Gemmell's. The *Engineering & Mining Journal* never specifically attacked the Jackling-Gemmell proposal, but the editors' basic point clearly applied: it would be impossible to make a profit mining low-grade copper ore under "the existing conditions in Utah." Anyone who suggested otherwise was probably an irresponsible if not downright dishonest speculator trying to lure the unwary into investing in a losing proposition.

The confidently dismissive editorial made an already questionable prospect look even riskier. As one prominent mining engineer later noted, probably ninety-nine out of a hundred engineers would have agreed with the journal's view at the time.[26] Most financiers, De Lamar among them, were troubled above all by the seven-figure investment Jackling would need up front.[27] Investing three million in start-up funds on a conventional ore deposit using well-known and proven mining technologies might be acceptable if the potential rewards were big. But it was a very different prospect to put three million dollars into a totally new type of deposit that would only be profitable if everything the engineers claimed about their revolutionary mining and milling techniques actually worked. The Jackling-Gemmell plan, for example, called for building a copper concentrator capable of processing three thousand tons per day. While some well-established mines with a proven steady supply of ore had concentrators that big or even bigger, most mining engineers would have considered even a five-hundred-ton concentrator to be very large for a new, unproven deposit.[28] The list of mining experts who declined offers to invest in the Bingham plan reads like a who's who of the industry: Ben Guggenheim, a member of the New York–based family of mining and smelting investors; Charles Coffin, president of Edison's General Electric Company; Marcus Daly, the head of the Anaconda in Butte; and William Clark, one of the early investors in Butte copper mining, to name some of the best known.[29]

Many years later, Jackling recalled that his plans for the Bingham mine "were subject to derision by mining engineers and engineering publications, and, in fact, characterized by some as reprehensible schemes of a get-rich-quick order, the promoters of which should be subject to punishment."[30] What Jackling himself did not fully appreciate, though, is that responsible mining financiers had every reason to be cautious of his Bingham proposal. Not only were Jackling and Gemmell proposing a new type of mining, they were also harbingers of a new model of the mining business. During the nineteenth century, the American mining industry had been chaotic and unpredictable, at times seeming more like a game of dice than a reliable business investment. Investors gambled on unknown deposits, hoping that as the mining proceeded it would discover more ore at depth. Many investors lost fortunes when promising deposits petered out to nothing, or processing technologies failed to live up to always optimistic expectations.[31] Still, the up-front investments were often not immense, and occasionally they paid off with extraordinary returns. The risks seemed warranted, at least to a certain breed of investor, variously described as bold or greedy, visionary or foolish.

By the time Jackling and Gemmell wrote their report, however, mining was on its way to becoming an increasingly stable long-term business, much more like the Saxon mines around Freiberg than the Wild West mines of California. The engineers had played a decisive role in this transformation, as had geology and the growing national and international dependence on cheap mineral supplies. As Jackling later said, "Mines are more often made than discovered."[32] Nowhere was this more true than with copper mining, where the supply of the increasingly critical industrial metal was repeatedly extended through ever more costly technological processes, including the pollution controls that kept environmental damages at least somewhat manageable. Jackling and Gemmell's new copper mining paradigm, however, was the most revolutionary yet. It brought to bear the forces of mass destruction that would have deep consequences not only for the mining industry but for other extractive industries as well, and hence for the entire material basis of modern industrial society.

In late 1899, however, these successes and their not always anticipated consequences were still many years off. Jackling and Gemmell's visionary plans for a gigantic open-pit mine in Bingham Canyon appeared to be dead in the water. No one who mattered believed they could make money mining 2 percent copper ore in an isolated canyon way out in Utah. Disappointed by De Lamar's lack of vision and the harsh criticism of the Bingham plan, Gemmell resigned. In time he would return to Bingham vindicated, but for now he headed south to work in less controversial Mexican mines.[33] Jackling was less able than his partner to let go of the Bingham prospect. As he worked on other projects during the next three years, he continued to search for investors, people with big enough imaginations and deep enough pockets to make the mine he had envisioned on paper a reality in rock. Perhaps he refused to let the mine go because he realized it was likely to be the only chance a former Missouri orphan boy would ever have to seize real fame and fortune.[34] Perhaps the sheer technological challenge was part of the attraction as well: the opportunity to apply on a vastly greater scale some of the same basic principles of measurement, control, and rationalization that had so intrigued him years before when watching a Missouri engineer survey a small-town street.

Whatever the explanation, Jackling never gave up on the Bingham prospect. Eventually, his persistence would pay off as other forces began to work in his favor.

THE "COPPER FAMINE"

On a Monday in the early spring of 1906, New Yorkers paging through their morning paper over coffee and toast could have found an intriguing headline on page 12 of the *New York Times*: "BOSTON COPPER GOSSIP —Metal Market Reported Bare of Copper." For those who cared about such things, this could have been either an alarming or a heartening head-line, depending on which side of the supply-and-demand seesaw they sat. Below, the news article related, "With the market pretty well cleaned up of copper and manufacturers hard put to secure needed supplies, mine own-ers see the day of 20 cents approaching." The *Times* reported that some ob-servers were speculating that this extravagant price of twenty cents per pound for copper might be a result of hoarding, while others blamed the scarcity on the lack of copper "refineries." But whatever the cause, supplies of copper were drying up, and manufacturers who depended on the metal were "unable to obtain enough for their needs, and must come into the market on the market's terms."[35]

The March *New York Times* article was a harbinger of turmoil in the copper markets to come. By the following year, some analysts were darkly speaking of an impending "copper famine." According to the mining engi-neer turned historian Ira Joralemon, by 1907 all the experts were confident that copper would never again "sell below twenty-five cents a pound as long as man used electricity."[36] For copper mining and smelting compa-nies, higher prices were obviously a blessing, so long as they did not go so high as to sharply reduce consumption. But for the manufacturers of elec-trical equipment and those who hoped for rapid national electrification (not to mention myriad other users of copper), the prospect of copper selling at twenty-five cents a pound was alarming indeed. Such a price would be nearly double what manufacturers had paid for copper over most of the past twenty years, and even just a year earlier the price had still remained below twenty cents. Cheap copper had helped spur American adoption of electrical power and lighting, and national consumption had nearly doubled since 1900 as a result.[37] Would high copper prices soon slow the pace of electrification?

No, as it turns out. By September 1907, the readers of the Sunday edition of the *New York Times* could have found a very different headline on page eight of their paper: "THE COPPER COLLAPSE."[38] Although the average price for copper in 1907 still worked out to twenty cents per pound, by the following year it was back down to thirteen cents. The dire (or optimistic) predictions of copper selling for twenty-five cents a pound would not come true until a decade later when the voracious appetites of a world war caused genuine worldwide shortages.[39] In just over a year, the "copper famine" had come and gone, and the metal was once again abundant and cheap in America. To some observers, it appeared as if there had never been a famine at all and the whole crisis had been manufactured by some mysterious unidentified cartel that had tried to corner the copper market.[40] There was, however, more to it than that. The experts predicting twenty-five-cents-per-pound copper for "as long as man used electricity" were not just foolish Cassandras.[41] The historical statistics for American copper production in the first decade of the twentieth century show there were good reasons to think the nation was heading for a copper shortage, if not a famine. From 1900 to 1906, the American output of copper had increased steadily from 291,000 metric tons to 489,000 metric tons—a jump of nearly 70 percent. During this same period, however, consumption had shot up by nearly 92 percent, suggesting the American output of copper was falling behind the exploding demand for the metal. Most telling of all, though, was the national output of copper in 1907: for the first time since the turn of the century, the total was actually down, dropping by about 4 percent to 468,000 metric tons.[42] Previously the growth of copper production had lagged behind the growth in demand, but now copper production was not just failing to keep up with demand—it was actually decreasing.

The fundamental cause for the copper shortage and increased prices was simple: the existing high-grade copper mines could no longer keep pace with demand. Worse, there seemed to be little prospect of discovering any new ones. As Joralemon argues in his 1934 survey of the industry, *Romantic Copper*, up until the early twentieth century the industry had kept up with the demand for copper by repeatedly discovering and developing rich new copper deposits. The Calumet and Hecla mines in Michigan's Upper Peninsula had met the modest demands of the mid-nineteenth century, yet they were much too small to have supplied cheap copper for early electrification.

In the 1870s and 1880s, though, Daly's development of the massive high-grade sulfide ores of the Anaconda mine in Butte, Montana, created a whole new source of copper. Thanks to the output from the Anaconda and other Butte mines, copper prices remained at about fifteen cents a pound into the early 1890s, despite rapid growth in demand.[43]

Even "the richest hill on earth," though, could not alone have kept pace with demand as electrification began in earnest. Joralemon gives credit for that accomplishment to the four big Arizona copper mining districts, all discovered between 1870 and 1880: Clifton, Globe, Bisbee, and Jerome. The Arizona deposits were different from those in either Michigan or Butte, both of which occurred as thick subterrestrial veins dipping and twisting through the rock. The desert deposits were almost as rich—as high as 20 percent copper in some areas—but were in immense lenses that often stretched out over two or three square miles. These new types of copper deposits and their remote desert location raised significant technological challenges, but engineers like James Colquhoun and James Douglas found solutions. Douglas's technical and managerial skill in particular helped turn Phelps Dodge, a modest copper trading firm, into one of the biggest copper mining enterprises in the nation. Phelps Dodge soon came to dominate Arizona the way the Anaconda did Montana, and subsequently developed copper mines around the world.[44]

Thanks to extraordinary combined output from the Butte and Arizona mines, by 1897 American copper production was eight times what it had been in 1880. Copper mining and processing companies actually feared they had been too successful and had created a copper glut. Indeed, for most of the 1890s copper sold for less than twelve cents a pound. Yet, as Joralemon argues, these painfully low prices were "the best thing that could have possibly happened to the copper miners."[45] Because copper was so cheap, it became the metal of choice for electrical supply grids, factory power systems, household wiring, and everything else associated with the American craze for electricity.

Butte, Clifton, Globe, Bisbee, and Jerome explain why it was even possible for national copper consumption to nearly double between 1900 and 1906. They also explain why the 4 percent drop in production between 1906 and 1907 had set alarm bells ringing among the mining engineers, geologists, and managers. These experts were reasonably confident there were no more Buttes left in the United States, nor any more rich sulfide copper lenses like those in the southwestern deserts. Likewise, the industry had only just begun to develop mines in other countries, and the copper deposits of South America and Africa were still largely unknown. Despite this, the national demand for copper continued to race blithely upward. Since all the rich copper deposits in the nation had been discovered, the only logical conclusion was that America was heading for a potentially disastrous copper famine.

Unless, of course, copper mines were not so much discovered as they were created—the point that Daniel Jackling had been trying to make back in 1899. For more than half a century, American copper mining had been like a giant national treasure hunt. The discovery of a fabulously rich ore deposit was the important thing. As long as there was enough high-grade copper in a deposit, the financiers tended to believe that engineers would find ways to get it out of the ground and purified. The Michigan deposits had ranged from pure 100 percent native copper down to very rich 10 percent ores; Butte's veins were between 50 percent and 12 percent copper; and the great Arizona lenses contained large masses of 20 percent copper. Whether the copper could be profitably mined, of course, had always depended on ore prices and available capital and technology. But by the standards of the time, all of these deposits were very rich deposits. Natural processes had done most of the work of concentrating the copper to the point where it could be economically mined.

The Bingham copper, by contrast, was very poor indeed: less than 2 percent copper at best. The Bingham deposit also looked completely different from anything the experts had seen before. Down in the Anaconda mine at Butte, miners could easily see the thick sheets and veins of copper as they chased them through the worthless country rock. Likewise, at Bisbee or Globe, anyone could spot where the big lenses of sulfide ore started and stopped. At Bingham, though, the primary deposit of low-grade copper ore had no obvious edges, no easily seen veins or sheets. Geological forces had concentrated the copper, but not by much compared to other mines of the day. Some sixty million years before, the upwelling of a gigantic plume of molten rock through a Mount Rainier–like volcano created the mass of copper at Bingham.[46] Before it had fully cooled, tectonic forces twisted and shattered the rock, creating hairline fissures and small cavities. Superheated water dissolved some of the copper in the original molten rock and forced it upward into the cracks above. When the water cooled, it deposited chalcopyrite (copper iron sulfide, $CuFeS_2$) and chalcocite (copper sulfide, Cu_2S) as tiny specks and stringers scattered throughout the mass of country rock. This secondary hydrothermal enrichment slightly concentrated the chalcopyrite and chalcocite crystals toward the center of the deposit, but there was no clear line or wall defining the extent of the deposit. Instead, the copper content was highest at the center and gradually declined moving outward. Measurable mineralization continued over a large area.[47]

Geologists call such deposits with flecks of minerals scattered throughout a mass of host rock a porphyry deposit. The term can be confusing. "Porphyry" originally referred to a specific purplish-brown igneous rock that the ancient Greeks and Romans valued for use in sculptures and architectural decorations. This stone had specks of larger crystals shot through a more smooth-grained rock matrix. Hence geologists later used the term to describe any rock with such a texture. The Bingham porphyry copper deposits are not at all purplish-brown. The granitic country rock is grayish-white, and the flecks of chalcopyrite and chalcocite within it reflect a brassy gold or metallic black sheen—or at least they do when visible. In the lower-grade Bingham ores, the crystals of chalcopyrite and chalcocite are so tiny and scattered they can be difficult to see with the naked eye.

Given these geological traits, it is easier to understand why many mining experts did not consider the Bingham deposit to be an ore deposit at all. Not only was the copper itself nearly invisible, the entire ore deposit had no clear beginning or end. The extent of the deposit was merely a mathematical abstraction. It could contract and expand from day to day, depending on the latest market price for copper or the success or failure of a new mining technology. More than any mine before it, Jackling's Bingham mine was obviously a cultural construction, a product of engineering and technology rather than a natural treasure that had been discovered. Whether it was an ore deposit at all or just a worthless mountain of rock depended on whether Jackling and Gemmell's careful measurements and experiments had been accurate and reliable. Human technology, planning, and organization would make Bingham a mine, if it was going to be a mine at all. Likewise, exactly where the "edges" of the Bingham ore deposit were depended on these same human factors. The limits to a Butte ore vein or an Arizona sulfide lens were obvious, but there were no obvious limits to the ore deposit at Bingham, and hence no obvious limits to how big the mine might become. Just where the ore stopped and the plain old rock began was not clear.

Jackling and Gemmell were thus proposing not only a new type of mine but a wholly new way of running the American mining industry. As Ira Joralemon later put it, success in modern mining would largely depend on planning and engineering, and "chance had very little to do with it."[48]

Mining financers were slow to understand this. In the four years since De Lamar had said no to investing in Bingham, Jackling had continued his search for a wealthy backer without success. Finally, in 1903 three Colorado Springs mining financiers—the brothers Spencer and R.A.F. Penrose and Charles M. MacNeill—agreed to fund mining on the property, but only at a scale much smaller than Gemmell and Jackling had envisioned. Between them, Jackling's three backers had a great deal of experience in mining engineering, geology, and finance.[49] Jackling's radical proposal made them understandably nervous. The financiers suggested Jackling begin with a modest experimental operation that would use conventional underground mining methods to extract the highest-grade ore in the deposit. There would be no massive steam shovels, no dedicated fifteen-mile railroad. Instead of Jackling's three-thousand-ton-per day concentrator, the partners agreed to finance only a modest three-hundred-ton experimental version.[50] All this would be only a shadow of what Jackling had envisioned in 1899, and he knew such a small operation could never profitably mine the millions of tons of lower-grade ore for long. For four years, though, he had failed to convince anyone to put up the big capital needed to build the mine at the scale he thought necessary. He would have to start small or he might never start at all.

On June 4, 1903, Jackling and his backers incorporated the Utah Copper Company. Soon after, the company purchased a controlling 80 percent interest in the Bingham hill property from its owner, Enos Wall. Jackling immediately began construction of the three-hundred-ton concentrating mill in the lower Bingham Canyon at an area that soon became the little town of Copperton. Meanwhile, a small mining crew began using a highly efficient underground mining technique known as block caving to "pick the eyes out of the mine"—that is, to remove only the highest grade ore possible. Jackling hired John McDonald, an associate from his Mercur mill days, to manage the underground mining operations. McDonald was an Englishman and had mastered block-caving techniques while working in the iron mines of Cumberland and Lancashire.[51] The key to the technique lay in carefully excavating a horizontal space underneath a large block of ore, supported by narrow pillars. When the pillars were blown, the ore would break and collapse under its own immense weight. Miners then had only to transport the broken ore to the surface. The small amount of ore mined during this early experimental period was nearly twice as rich in grade as the typical ore Jackling mined fifteen years later.[52] By mining only the best ore in the property, Jackling's new Utah Copper Company thus managed to pay dividends to stockholders after only one year of opera-tions.[53]

The fledgling little Utah Copper Company seemed to be a success. But Jackling knew that it was all an illusion: small-scale concentrating and un-derground mining would fail as soon as the limited higher-grade deposits gave out. During the first year of experimental operations, he had contin-ued to refine his plans for an open-pit operation combining steam shov-els, a railroad, and a big concentrator.[54] To make it a reality, he still needed a lot more money than his Colorado backers were willing or able to put up. Fortunately, the impending fears of a "copper famine" were now beginning to work in his favor. Part of the reason the Utah Copper Company had been able to pay a dividend in 1905 was that copper prices were already rapidly climbing. When Jackling completed the Copperton concentrator in 1904, copper sold for thirteen cents a pound. By 1905 it was up to sixteen cents, and in 1906 it would be at nineteen cents.[55] The improving price and the emerging possibility of a national shortage made investing in any copper mining proposal much more attractive—even one as revolutionary as Jackling's.

One obvious source of big money was the General Electric Company, the enterprise that had evolved out of Thomas Edison's early Edison Gen-eral Electric Company. Eager to assure itself of an adequate supply of cop-per for its production of wire, generators, and other electrical devices, the company was already contemplating a move into copper mining. None-theless, following a careful inspection of the Bingham property, the board of directors declined. They reportedly found the huge estimates of poten-tial copper production from the mine simply too hard to believe.[56]

In late 1905, Jackling finally found his visionary partner. The Guggen-heim Exploration Company, a subsidiary of the wealthy Guggenheim mining and smelting conglomerate, agreed to provide nearly eight million dollars in bond and stock financing for Utah Copper. In return, the Gug-genheims gained a quarter interest in Utah Copper as well as a lucrative twenty-year contract requiring the company to sell all of its copper ore concentrates to the family's smelting branch, the powerful American Smelt-ing and Refining Company, or ASARCO. Shortly thereafter, ASARCO began construction of a big new smelter near the banks of the Great Salt Lake. Only two miles north of Jackling's concentrator at Copperton, ASARCO's Garfield smelter would process the vast amounts of ore pouring out of the Bingham mine for decades to come. The ASARCO engineers took as their model the Anaconda's Washoe plant that had opened in early 1902 with

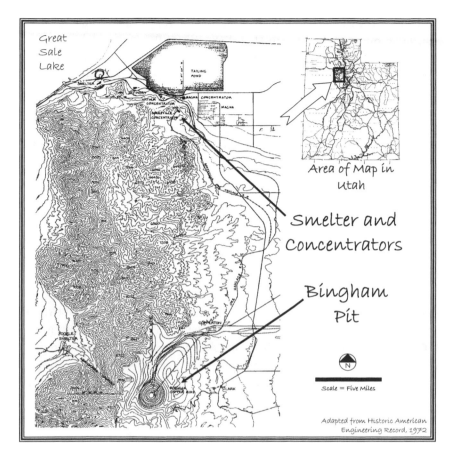

Great
Salt
Lake

Area of Map in
Utah

Smelter and
Concentrators

Bingham
Pit

Scale = Five Miles

Adapted from Historic American
Engineering Record, 1972

11. A contemporary map of Jackling's Bingham Canyon mining complex south of Salt Lake City. Jackling transported the ore via rail to the Arthur and Magna mills, where concentrating machines removed the tiny amounts of copper and sent the waste to the tailings ponds. The ASARCO smelter on the banks of the Great Salt Lake drove off the remaining impurities, releasing some of it as pollution into the air of the Salt Lake Valley. *Adapted from Library of Congress, Prints and Photographs Division, Historic American Engineering Record, Reproduction Number HAER UTAH, 18-BINCA, 1-2.*

such disastrous consequences for the Deer Lodge Valley. The Garfield smelter, though, would eventually eclipse the Washoe to become the largest in the world.[57] As we have seen, it would also soon face pollution suits that threatened to shut it down. Thanks to Frederick Cottrell's electrical precipitator and other technologies, however, it survived to continue processing the ore pouring out of Jackling's mine and concentrator.

With the deep financial resources of the Guggenheim family now at his disposal, Jackling began making the giant open-pit mining operation he and Gemmell had envisioned in 1899 a reality. By 1908, eight steam shovels would be hard at work chewing into the Bingham hill and his new Magna concentrator would be one of the biggest in the world. It was all there just as planned, though the operation would soon become even bigger and more destructive than the two ambitious young engineers could have even imagined in 1899. The process of making a mountain into a molehill had really only just begun.

Not in the least coincidentally, the year 1908 also saw the price of copper drop to thirteen cents a pound, down 35 percent from its high of twenty cents the previous year. Copper would not sell for above twenty cents again until World War I. Just as the Butte and Arizona deposits had done in previous decades, the Bingham deposit provided a vital new supply of copper to the nation, guaranteeing that the metal would remain abundant and cheap. The Bingham hill of copper that so many had insisted was worthless would inject more than a hundred billion pounds of new copper into the world economy in the decades to come. This was more copper, Ira Joralemon pointed out in 1934, than humans had "used in all the ages since the first cave man picked up the first copper pebble."[58]

Commentators, both then and now, have all agreed that Jackling's Bingham mine was a remarkable technological and managerial achievement. Many have misunderstood, however, both the nature of what Jackling had accomplished and the broader consequences of his revolutionary ideas. Even as keen an observer as Joralemon believed that Jackling's "contribution to the world" was really just a simple matter of economies of scale. Jackling's fixed costs, Joralemon argued, were largely independent of the amount of ore treated. If Jackling could divide his costs by the profits from processing ten thousand tons of ore instead of a thousand tons, the expense per ton would be minimized and mining made profitable.[59] Joralemon's mistake is, in retrospect, understandable. Others had made the same error in trying to explain how mass production succeeded in making cheap automobiles and refrigerators. Of course, scale was important, but equally significant was the related matter of *speed*, or what the brilliant business historian Alfred Chandler later called "throughput." Just being big was not enough. The real trick lay in being *fast*.

In this need for speed lay the roots of mass destruction.

MASS DESTRUCTION

Daniel Jackling never expressed much appreciation for the beauties of the natural world—at least not in the record left in his published and unpublished comments. Back in the 1870s, Rossiter Raymond had been transported by the "exquisite rose-tinted Abendroth" on the rugged peaks of the Wasatch Mountains west of Salt Lake City.[60] Jackling could have easily viewed those very same Wasatch peaks from the flanks of the Oquirrh Mountains where he had begun excavating his giant pit, and perhaps he did stop at times during his busy days to appreciate the sight. Still, it is difficult to imagine Jackling penning the kind of poetic natural appreciations of a mountain peak written by Rossiter Raymond. One suspects he would have found the romantic nature worship of such men as Raymond or Clarence King to be embarrassingly feminine and soft-headed, inappropriate to the hard-edged rationality needed to manage a modern industrial mine. Certainly, Jackling never fancied himself an engineer-poet like those men. He was preeminently an engineer-businessman, and he showed little interest in the humanistic matters that had inspired many in the first generation of elite mining engineers. Even his new open-pit mining technology stripped away the romance that Raymond and others had once found in the shadowy world of the subterrestrial mine. Whatever mysteries may have been hidden in those long, dark tunnels of the underground mine would now be laid bare.

Jackling was a new type of mining engineer appropriate to a new type of mining. By the early twentieth century, the elite first generation of American mining engineers who had trained at the top schools in Germany or France was passing. Raymond still held tenaciously to his control of the American Institute of Mining Engineers, but he would step down in 1911 and would die seven years later. John Hays Hammond was still the most famous mining engineer in the world, and he played an early role at Bingham as the Guggenheims' consulting engineer and manager, but by 1907 he was in ill health, and thereafter his interests increasingly shifted to politics and international affairs.[61] The graduates of the top American mining engineering programs at Columbia and MIT constituted a new home-grown elite within the profession. Increasingly, though, these were not the engineer-humanists of the past but narrow technological specialists with considerably more training in business management than in the classics. The new generation of mining engineers still had men with a literary bent, like Ira Joralemon. Tellingly, though, when Joralemon first published his book on the history of world and American copper mining in 1934, he still felt he could title it (somewhat nostalgically even then) *Romantic Copper*. When he updated the book in 1973 with six new chapters, he instead called it *Copper: The Encompassing Story of Mankind's First Metal*. As he notes in the introduction to the new edition, the earlier name no longer seemed fitting since "this sort of romance has gone out of style."[62]

And so it had. If the earlier generation of mining engineers had believed they could use science and technology to minimize natural damage and harmonize mining and nature to some degree, Jackling's destructive new mining method made such a balance far more difficult to achieve. Jackling occasionally praised conservation, but only as a means of maximizing the extraction of the nation's natural resources, not as a means of minimizing natural degradation. He had no sympathy for "so-called" conservationists who would allow river water to "run away to the sea and be wasted" instead of being harnessed for hydropower, and he considered it sheer idiocy to let any mineral deposit sit idly undeveloped when the technology for its economic exploitation was at hand.[63] As John Hays Hammond later noted, one of the accomplishments of Jackling and other creators of the low-grade mining technologies was that they essentially eliminated any threat of a copper shortage for several decades to come. Mining engineers like Jackling, Hammond argued in 1933, were now able "to furnish copper in such quantities and at such rates as the world may need."[64] Showing little interest in the preservation of an idealized nature for its aesthetic or spiritual powers, Jackling saw conservation simply as a means of maximizing the copper available to the world while also maximizing profits to shareholders. Others, including engineers and scientists like Frederick Cottrell, would continue their efforts to balance mining with the natural world, but the future was increasingly with Jackling and his stunningly destructive pit.

Any man who would propose the godlike task of reducing an entire mountain to dust in order to grasp the tiny specks of copper in its rock would likely need more than the typical human allotment of hubris. Undeterred by any romantic affection for wild nature, buoyed by the confident spirit of a modern new century, confident in his technological prowess, Jackling had enough to keep his vision alive until the deep well of capital offered by the Guggenheim fortunes made its realization a possibility. Jackling was not alone in making the pit a reality. Dozens of engineers, inventors, and managers contributed.[65] Jackling also provided few technical innovations; instead, he brought together existing technologies to create something entirely new.[66] Others had successfully processed low-grade copper ore, though always as an adjunct to the mining of richer ores.[67] Others had used steam shovels, but never on such a large scale. Others had developed open pits, but not to mine an entire mountain of copper ore. Jackling's accomplishment was to bring all these together in a way that redefined the very meaning of what constituted a "mine."

Jackling's technology of mass destruction shared key traits with the mass production manufacturing technologies Henry Ford and others were perfecting in their factories during precisely the same period. Both Ford and Jackling founded their companies in the same year, 1903, and the two men drew on similar principles to increase speed and drive down costs. The result in Ford's case was astonishingly cheap automobiles, and in Jackling's astonishingly cheap copper. Indeed, since Ford's autos used copper in their radiators, carburetors, and electrical systems, cheap copper helped make cheap autos. When Ford began mass-producing his new Model T, he actually increased the amount of copper in each car by five pounds.[68] Given these and other parallels, commentators both then and since have often referred to the Bingham mine as a form of mass production mining.[69] As the mining engineer and historian A. B. Parsons wrote in 1933, the development of large-scale copper mining was "an outstanding example of the advantages of mass production coupled with the fruits of scientific research and inventive genius."[70]

Too simple a comparison to the mass production technology used by Ford and others, however, can obscure deeper insights into exactly what Jackling had wrought. What similarities there are reflect a convergence of ideas rather than any direct borrowing or the effect of remote influences. Jackling did not apply the principles of mass production to mining—there were no manuals of mass production techniques for him to consult at the time, and his approach was developing at roughly the same time as Ford's and others in manufacturing industries. Indeed, as the historian of technology David Hounshell notes, the term "mass production" was not even coined until the 1920s.[71] More important, Jackling created a system that was distinctly different from mass production in several critical ways. Hence it deserves the related but much more accurate term "mass destruction." The name itself suggests one of the more obvious differences between the two systems. Mass production factories brought together various pieces of raw material from the natural world in order to produce goods as rapidly and efficiently as possible. In mass destruction precisely the opposite occurred. Jackling's Bingham mine took apart the natural world in order to produce raw materials as rapidly and efficiently as possible. Both ultimately resulted in the production of some valued good, whether that be a Model T or pure copper metal, but only Jackling's system first demanded the wholesale destruction of part of the natural world. Even the famous *dis*assembly lines at Chicago's meatpacking plants, which were important precursors to Ford's assembly lines, were very different from mass destruction. Chicago's meatpacking workers required considerable skill and precision to cut apart the cattle and hog carcasses moving by them on mechanized conveyors. A process analogous to mass destruction would have entailed grinding the entire carcass into tiny bits and then using specialized machinery to separate the valued meat from the various other useful waste products such as bone, hooves, and the like.

Imagining such a grisly but potentially efficient mean of extracting meat from animals points to another key difference between mass destruction and mass production. Whether the process entails assembling cars or disassembling cattle, mass production has typically taken place in a factory, a discrete human-built structure typically viewed as distinct and separate from the natural world. Indeed, from the time of the earliest industrial factories in Great Britain, humans have tended to see factories as the antithesis of the natural world, the very epitome of the artificial and human-made. Frequently constructed in densely populated urban areas, factories brought together large numbers of workers to labor in tight coordination with specialized machines driven by hydropower or hydrocarbon engines. Though the builders of some early American textile factories, like the famous Lowell Mills of Massachusetts, tried to suggest their factories were in harmony with nature, the opposite view quickly won out. The American factory, like William Blake's "dark satanic mills" in England, seemed almost to define the technological, artificial, and unnatural in the popular imagination.

This traditional concept of the factory has been a serious impediment to fully understanding the significance of Jackling's system of mass destruction. Few have understood that Jackling's mine was itself an immense natural factory of mass destruction. This key insight has eluded most commentators because they have tended to embrace the conventional idea of the factory as a discrete structure that is both literally and metaphorically walled off from the natural world. However, the mass destruction factory is not separate from the natural world—the mass destruction factory *is* the natural world. In Jackling's system, nature itself was a factory carved out of natural stone and intimately associated with the fused environmental and technological systems used to create and operate it.

Again, the mine had always been a part of nature, but the Western tendency to see mining as a distinctly unnatural pursuit has often obscured this reality. Most have tended to assume that the subterrestrial mine was as seemingly unnatural and utterly anthropomorphic as Henry Ford's Dearborn, Michigan, assembly plant. But to do so is to make a serious analytical and conceptual error. As argued earlier in this book, although the work of some engineers took place in largely human-made settings, mining engineers had always worked with and in the natural world. They were, in this sense, among the first environmental engineers, second perhaps only to the civil engineers. For centuries, mining experts had used increasingly sophisticated technology to engineer subterrestrial environments in which biological creatures such as human beings and animals could survive and work. Like a factory, these mines were human-made. But like a farm or a managed forest, they also remained a part of the natural world. Mining engineers thus had to understand and work with the natural conditions they encountered in the mining environment, as well as attempt to control or modify these conditions to make biological life in the mines sustainable. A chemical or electrical engineer might create an urban factory out of whole cloth, building it from imported materials on the edge of a city where it appeared to be (though in fact was not) completely separated from nature. The mining engineer's creation always remained clearly and very literally rooted in the natural world. The subterrestrial mine was both natural and artificial, both ecological and technological, both an environment and a factory.

In recent years, a number of perceptive scholars have begun to suggest the need to redefine what we mean by a factory. The roots of much of this innovative new work can be traced back to Richard White's seminal little 1995 book *The Organic Machine*, which suggests the Columbia River is best understood as a hybrid technological system that mixes human and natural forces. White does not argue that the Columbia River constitutes a type of factory. But by demonstrating that the traditional cultural lines drawn between anthropomorphic and natural systems are often porous and indistinct, White raises the possibility that a "machine" might have both environmental and technological components.[72] Drawing on these insights, the historian David Igler argues that large-scale ranching techniques in late nineteenth-century California created a type of factory in which the natural environment was part of the technological system. In the fittingly titled book *Industrialized Cowboys*, Igler shows how California cattle barons used the varied environment of their vast landholdings as part of their factory system, moving their cattle to different ecosystems to compensate for changes in season and weather. As in a traditional enclosed factory, this outdoor "natural" factory was designed to maximize production—in this case the production of meat by highly bred and "engineered" animals.[73] Indeed, this idea that even some animals might usefully be seen as highly engineered organic technologies has been most brilliantly argued by the historian of technology and the environment Edmund P. Russell. In several recent seminal pieces, Russell suggests that only a traditional and largely inaccurate tendency to see technologies as inorganic mechanical phenomena has kept us from understanding that some animals, crops, and other organic entities are, at least in part, anthropomorphic technologies.[74] Of course, such organic technologies are not entirely human creations; like mines, they are hybrids that combine the properties of natural and artificial, ecological and technological.

If cows, wheat strains, or bacteria are technologies operating within complex hybrid envirotechnical systems, then clearly the old concept of the factory as the antithesis of the natural and organic no longer makes sense. Another leading historian of technology and the environment, Deborah Fitzgerald, makes a similar point for farming. In *Every Farm a Factory*, Fitzgerald shows that in the late nineteenth century American farmers began to practice a type of industrialized farming whose principles were similar to those of urban factories. The organic and ecological elements of the factory farm were closely integrated with hydrocarbon-powered machines like tractors and threshers in order to increase crop yields and speed harvesting and processing.[75] Clearly, the farms Fitzgerald describes are species of factories, but they are factories constructed in and out of the natural environment itself, not some brick and iron building seemingly far removed from nature.

Given this emerging new understanding of factories, it becomes much easier to recognize the subterrestrial mine as a type of factory that combines environmental and technological systems. Even more important, this environmentally based definition of a factory leads to a critical insight: any change in the way the subterrestrial mining factory functioned also generally necessitated a change in the environment, as the two could not be functionally or logically distinguished. The same is true of other envirotechnical factories, like cattle ranches and industrial farms. This is a critical insight that the old idea of factories as entities apart and isolated from nature tended to obscure. Adopting mass production techniques at Ford's Dearborn plant likely had environmental consequences, but they were difficult to see without tracing raw material supply chains backward or waste disposal and product use chains forward. In a more obviously envirotechnical factory, however, the consequences of adopting mass production techniques were immediate and apparent, as they necessitated a fundamental change in the interface between the technological and environmental systems.

At a trivial level, any increase in production in a mine through, say, a more powerful hoist or better pumps had always meant more ore would be removed from the underground environment. The mine would grow bigger, so long as reasonably high-grade ore remained. Ultimately, though, the subterrestrial environment itself presented a serious obstacle to continued increases in the size of a mine and the speed of extraction. The deeper a mine went, the higher the costs for taking water and ore out of the mine and sending men and air in. The confined space and air of the underground environment also sharply limited the kinds of machinery engineers could use there. Likewise, mine shafts were literal bottlenecks, sharply restricting the speed at which material could flow in and out of the mine. If mining was to make full use of the mass production principles of high throughput, then the structure of the factory—and thus of the environment—would have to be radically changed.

Put simply, mass production mining entailed mass destruction of the environment itself.

For reasons that will become clear, the specific environment of Bingham combined with the mining technologies and cultural beliefs of the day so that open pits appeared to be the most "logical" means of increasing speed. Such clear environmental consequences were less obvious when adopting mass production in a conventional factory, though the contrast fades if we view urban factories as technological systems deeply linked to the environment. Indeed, rather than mistakenly seeing envirotechnical factories like Bingham as akin to our conventional idea of a factory, we would do better to realize that all factories are envirotechnical entities akin to the Bingham Mine. At least Jackling's pit of mass destruction had the virtue of being obvious about its environmental consequences. As such, it was and continues to be a powerful symbol of a much broader relationship between modern industrial society and the natural world—a theme we will return to in the final chapters.

For all the reasons mentioned here, recognizing this reality has never been easy or obvious, as it requires that we break down deeply held cultural beliefs about clear lines between the natural and artificial, the environmental and technological. To an earlier generation of American mining engineers, even the largest mine had once seemed insignificant in comparison to the vast expanse of the natural world around it. To them, the technological sphere was only one small part of a much larger and still largely dominant natural sphere. Mining took place as a part of the natural world—to echo the cultural historian Leo Marx, the machine was still in the garden and part of the garden.[76] Romantically minded engineers like Raymond and Rickard could still reasonably hope to create mines that harmonized, at least to some degree, with the natural world.

With Jackling's pit, though, the technological threatened to overwhelm the natural, to become so pervasive as seemingly to constitute a human-made realm entirely apart from nature. Indeed, the envirotechnical factory that had previously been confined and hidden underground now came to the surface, its full potential for reordering the face of the planet made devastatingly obvious. Minimizing the environmental damage of such a mine was, in at least some sense, a logical impossibility. Engineers could and would strive to limit the environmental pollution from concentrators, smelters, and other parts of the mass destruction system. Yet it obviously made little sense to suggest that better engineering and technology could somehow "conserve," much less "preserve," the old Bingham hill. The hill was no longer a hill, it was the pit, and the abstract line that separated the edge of the pit and the rest of the planet was steadily expanding—or perhaps even fading away all together. The machine was no longer in the garden—the machine quite literally was the garden.

The sheer destructiveness of Jackling's mining system was never his goal, of course. His goal was to make money through the cheap and efficient disassembly of the environment in order to extract the tiny percentage of copper that the society of the day considered valuable. As in conventional mass production, Jackling relied on the economies of scale and speed to increase copper output and drive down the cost per unit. To do so depended largely on replacing human and animal energy with crude but far more powerful machines driven by hydrocarbons or hydropower. The mass destruction system had three essential stages, each of which had environmental consequences. The first stage was ore breaking, the task of shattering the hard-rock ore using the chemical power of nitroglycerin, now in the tamed and refined form Alfred Nobel called dynamite. Guiding this precise chemically powered reduction of mountain to rubble was the radically improved ability of engineers to "see" underground via new remote drilling and sampling techniques. In the second stage, the ore had to be picked up, loaded, and moved as quickly as possible. Here speed was of the essence, and steam shovels and trains greatly accelerated the process through the injection of immense amounts of artificial power. In the third and final stage, improved means of crushing the rock to dust and separating out the tiny amounts of copper were key. These milling and concentrating technologies made it possible for the first two stages to exist by replacing the selective mining of skilled and costly underground workers with the nonselective and energy-intensive mining of crude steam shovels.

All three stages were intimately linked. None could function without the others, and a good deal of Jackling's accomplishment was his merging and refining of these three parts into a seamlessly efficient whole. The result was a hybrid indoor-outdoor factory designed to chew up an immense chunk of the earth's crust, extract the tiny percentage of desired metal, and spit back the remainder as equally immense volumes of waste.[77] In clearly explaining the logic of Jackling's system, there is always a danger that it will begin to seem inevitable, as if it were the only possible way to mine. This, however, is as much an illusion as the culturally constructed separations between the technological and environmental. Mass destruction did make sense, of course, but only in the context of a particular and peculiar set of economic, ideological, and political beliefs that dominated in America in the twentieth century. To paraphrase the contemporary actor and humorist Alan Alda, Jackling's mine ran like clockwork—unfortunately, it was quite arguably a cuckoo clock.

Fittingly for a system of mass destruction, the first step was to blow it all up.

DYNAMITE: THE ART OF PRECISION
CHEMICAL DESTRUCTION

In the early 1920s, the DuPont-owned Hercules Powder Company ran a series of historically based advertisements on the theme of the human mastery of the natural world. One, entitled "The Pyramids of Gizeh," pictures Egyptian slaves straining to pull a block of limestone up to a rising pyramid while an overseer goads them onward with his whip. "Herodotus records that 100,000 workmen toiled for a generation to build the great Pyramids of Gizeh," the ad notes, but in the current era "explosives have replaced slave labor and have made possible the economical production of ore, coal, and rock products required by modern civilization."[78]

Explosives manufacturers often turned to the theme of ancient engineering works to advertise the power of their modern products.[79] Comparisons with the past were useful ways to suggest that modern technology in general, and explosives in particular, had allowed humans to realize the age-old dream of subduing nature. Likewise, there was no longer any need for cruel overseers with their sharp-barbed whips. The enslavement of thousands of people to create what Lewis Mumford called the "megamachine" of the past was no longer necessary. Mumford estimated the megamachine that built the Great Pyramid generated about 2,500 horsepower, and the assembled laborers "collectively performed the equivalent of a whole corps of power shovels, bulldozers, tractors, mechanical saws, and pneumatic drills." [80] Now Hercules suggested that a single stick of dynamite could do the work of dozens of enslaved men and women, while a few cases of these sticks might do the work of thousands. When the DuPont company collected the advertisements in a 1923 book, the title captured the spirit of the campaigns and the centrality of dynamite in large-scale earth-moving projects: *Conquering the Earth*.[81]

DuPont's focus on the sheer destructive power of dynamite, however, is a bit misleading. Oddly enough, the finesse and precision provided by dynamite were nearly as important as its greater destructive power in enabling Jackling's system of mass destruction to work. Before dynamite, miners had relied on the ancient explosive mixture known as black powder. At the end of the eighteenth century, the famous French scientist Antoine Lavoisier had developed a sophisticated understanding of black powder and its manufacture. He passed on his technical expertise to his student E. I. du Pont, who migrated shortly after to the United States, where he established the American black powder industry.[82] A century later the DuPont company had developed manufacturing procedures to produce the best black powder possible, yet the formula remained essentially identical to that of centuries before: a thoroughly incorporated mixture of saltpeter (potassium nitrate), charcoal, and sulfur. Still, this simple mixture sufficed to power the early American boom in mining and construction that began during the middle of the nineteenth century. As a 1928 article in the *DuPont Magazine* boasted, echoing elements of Frederick Jackson Turner's famous frontier thesis, "civilization went West sitting on a powder keg."[83]

For most mining and earthmoving operations, black powder worked reasonably well. But when Jackling developed the Bingham open-pit mining system, he ultimately rejected black powder in favor of the latest type of nitroglycerin-based explosives.[84] Invented by the Italian chemist Ascanio Sobrero in 1847, nitroglycerin is a powerful but frighteningly unstable explosive. Most engineers and managers considered it much too dangerous for practical use. The Swedish businessman Alfred Nobel disagreed and began manufacturing and marketing the substance near Stockholm in 1861. Nobel was either blind to the dangers of his "patent blasting oil" or willfully ignored them, and his early attempts to market nitroglycerin were plagued by a completely foreseeable series of spectacular accidents and frequent deaths. Many nations soon banned the manufacture and use of Nobel's blasting oil. But in the United States some blasters were so impressed by the far greater power of nitroglycerin in comparison to black powder that they continued to use it. In tunneling through hard rock, nitroglycerin could often double the rate of progress. Still, nitroglycerin would have remained a hazardous explosive with only a limited market but for Nobel's ultimate success in taming his blasting oil. By mixing the nitroglycerin with a type of soil known as *Kieselghur*, Nobel was able to stabilize the explosive while preserving its power. He called his invention dynamite.

Nobel helped found the first American dynamite company in San Francisco in 1877. The company did a brisk business supplying western gold and silver mines. Soon after, an old black powder manufacturing company called the California Powder Works improved on the Nobel dynamite formula by using a stabilizing mixer that also added to the explosive force. Recalling the strength of the mythic Greek hero, the company called its new explosive Hercules. The new type of dynamite proved so successful that the California Powder Works eventually changed its name and became Hercules as well.[85] Meanwhile, the East Coast–based DuPont company continued with its focus on manufacturing black powder for several more years while warily eyeing the growing demand for dynamite. In 1881, DuPont moved to meet the competition by buying the Cleveland branch of the Hercules Powder Company, and a year later Lammot du Pont founded a new high explosives company called Repauno. Soon DuPont became a major player in promoting the use of dynamite in mining and other fields.[86]

Recalling the grisly and widely publicized stories of earlier nitroglycerin accidents, many miners initially resisted the introduction of dynamite. But gradually the undeniably superior power (and hence potential for increased profits) of dynamite won most of them over. As the Joplin, Missouri, *Daily News* reported in 1878, "After many trials and much opposition it has been proved to be far superior to any other, and today is the favorite explosive of all miners in this region who make it a maxim to use only that which is the best, cheapest and safest."[87] The superiority of dynamite in hard-rock blasting largely stemmed from the rapidity of its explosion, which was much quicker than that of black powder. The force of a nitroglycerin explosion is like a sharp, rapid hammer blow in comparison to the much slower and (if the word does not seem too bizarre) "gentler" explosion of black powder. Explosives experts call this property the degree of brissance. A high-brissance explosive like nitroglycerin could shatter rock that black powder merely lifted but did not break. For the hard-rock ores in which copper and silver typically lay, this shattering effect was an indispensable time and labor saver. In a soft mineral like coal, though, the increased brissance of dynamite tended to pulverize it into dust—a serious problem when most coal mine operators wanted fist-sized pieces. Dynamite manufacturers solved the problem by varying the amount of nitroglycerin they added to the absorbent material. This produced a range of dynamites with strengths from 5 to 60 percent of pure nitroglycerin and allowed manufacturers to sell the less powerful dynamites more cheaply.

Fine-tuning the power and brissance was only the beginning. Soon DuPont and other dynamite makers began to offer an array of different special-purpose explosives. With black powder, a mine operator had been able to choose from a handful of products differentiated by powder grain sizes and small variations in formula. But by 1913, explosives companies were advertising several thousand different varieties of high power products. DuPont's Repauno plant alone offered 914 different high explosives.[88] Some of these variations were so-called permissible explosives, designed to be used safely in gas-filled coal mines. Some were made using ammonia nitrate in the mix, a chemical with explosive power similar to nitroglycerin's but cheaper to make and more stable. Some were low-freezing-point dynamites for use in northern regions, since straight dynamite froze at forty-six degrees and then required a dangerous thawing process. And in a move that was especially important to the development of open-pit mining, dynamite manufacturers also began to offer explosives with a broad range of strength versus speed of explosion. As one commentator noted, manufacturers succeeded in making "small changes that have resulted in a variety of explosives which yield any desired type of fragmentation and are readily adapted to specific conditions."[89] Assessing the three factors of speed, strength, and cost, smart blasters could select the most effective and economical explosive for shattering their specific rock in fragments of the desired size. In other words, the explosive was carefully matched to the natural environment in which it would be used.

Dynamite and the other new high explosives reached their greatest level of finesse and control under the guidance of Charles L. Patterson, a pioneering Repauno dynamite salesman who was among the first in the industry to advocate the careful matching of explosives to the blasting environment. Repauno inaugurated a research and development department to create both variations on established formulas and entirely new types of explosives.[90] At the same time, the Repauno sales force became the liaison between the explosive user and the research department. One customer, for example, needed a gelatin explosive for use underwater; at the request of the sales department, the Repauno lab developed and field-tested a custom-made product.[91] Explosives salesmen also provided technical expertise to mine operators on which explosive would work best for their rock, while company catalogs and pamphlets offered similar information on their safe and efficient use.

In time, Jackling's Bingham mine became one of the most sophisticated users of precision dynamite blasting in the West. During the early days, though, Jackling first used conventional black powder for blasting. The logic of this was a result of the unique history and topography of the Bingham hill. Earlier underground miners had already extensively tunneled into the hill in pursuit of the relatively small amounts of higher-grade copper ores. These hollow spaces in the rock interfered with the usual method of drilling holes that were then packed with dynamite. Instead, Jackling used a technique miners called "gopher blasting," which took advantage of the ready-made cavities in the rock to drill further tunnels about thirty feet behind the face of the hill, or the "bank" in open-pit mining terms. Miners filled the tunnels with huge amounts of black powder, sometimes as much as eight hundred thousand pounds, and then sealed the tunnel entrances with concrete bulkheads. When the mass was ignited, the resulting explosions could be awesome in their earthshaking power. Some gopher blasts broke more than a million tons of rock in a single immense explosion.[92]

Though cheap and powerful, gopher blasting with black powder could not provide the sort of precision shattering of the rock that Jackling increasingly demanded for maximum speed and efficiency. Once shattered by the explosion, the rock had to be lifted and loaded into railcars by a power shovel, and power shovels could move finely fragmented ore at three or four times the rate it could move uneven, "bouldery" rock. A boulder that was too big to fit into a shovel dipper (the big lifting and digging part of the shovel) had to be blasted again by the explosives pit crew, squandering both time and money. Alternatively, the power shovel operators could break the boulder into smaller pieces by pounding it with the dipper, but this also wasted time and led to rapid wear and frequent breakdowns of the expensive shovels.[93]

Given the importance of speed and cost savings to the profitable operation of the Bingham mine, Jackling increasingly switched over to dynamite within the first few years. As the operations pushed further into the hill, the miners also encountered fewer old tunnels, which enabled them to use power drills to carve holes in the rock for dynamite. Though the methods would change somewhat once the hill became a pit in the 1930s, during the first decade of operations Jackling perfected the basic procedure for blasting a hill into pieces. Whether excavating a hill or a hole, the two essential units of any open-pit mine are the bench and the bank. The bench is the flat terrace on which workers, shovels, and trains move. The bank is the vertically sloping face between the benches. When Bingham was still a hill, the series of benches and banks gave the mine the appearance of a stepped pyramid, though the steps are much larger than any structures found in Egypt or the Americas. The Bingham benches varied between fifty and one hundred feet in width, and the banks were forty to seventy feet tall, the equivalent of a three- to five-story office building. Drilling and blasting occurred in three stages. First, drilling crews used percussion air-powered drills to drive twenty-two-foot horizontal "toe holes," so called because they were drilled into the base of the bank. Second, the blasting crew packed the holes with dynamite. The crew then cleared the area of men and equipment and detonated the charges, resulting in a powerful blast and an avalanche of rock cascading down from the face. Third, the crew returned to drill a second series of holes along the face, this time from more precarious positions atop the debris halfway up the bank, and these were again charged with dynamite. The crew did not detonate the second holes until after the steam shovel had removed the rock from the first blast. The shovel then retreated, the second charges were ignited, and the shovel returned to load the remaining ore onto railcars.[94]

Jackling's two-stage blasting technique might seem a bit cumbersome, and rightly so. As the pioneering experimenter, Jackling had made a mistake in adopting seventy-foot banks that could only be broken by two explosions. Subsequent open-pit mines typically had forty- to fifty-foot-tall banks that could be shattered with only one shot. These shorter banks were also less dangerous for workers and machinery, in the not unheard-of event that a shovel or steam engine happened to tumble off an edge. Having begun the excavation of the hill with seventy-foot banks, though, Jackling was essentially committed to them for the next few decades. As new benches began and the pit replaced the hill, the shorter heights were adopted.[95]

Blasting was by far the most expensive operation in Jackling's system. Unlike other processes in the mine, such as shoveling and transport, the work of the drilling and blasting crews could not be easily mechanized. Even as late as 1938, Utah Copper still had a crew of 555 men blasting every day. Jackling and his engineers eventually settled on one type of dynamite as the most effective for the Bingham rock: Hercules Red H No. 3, a 60 percent ammonia nitrate dynamite made by the DuPont-owned Hercules Powder Company.[96] The dynamite actually contained only 10 percent nitroglycerin, while the remaining explosive power came from 80 percent ammonia nitrate, but the effective combined power was 60 percent of that produced by a pure nitroglycerin explosive.[97] The Bingham mine's appetite for ammonia nitrate dynamite was so voracious that Hercules built a manufacturing plant in the neighboring town of Bacchus, Utah.[98] As the mine grew and the ability of the shovels to handle more ore per shift increased, the typical size of the Bingham's earth-shattering explosions also kept pace. By 1918, Bingham explosions sometimes broke as much as twenty-five thousand tons of ore with a single blast. During one month in 1918, workers detonated 369,050 pounds of dynamite in order to break just over a million cubic yards of solid rock.[99]

Jackling did not play a direct role in the development of dynamite and the wide array of modern explosives that followed. However, the mining system he created depended on the increased precision and efficiency of the explosives paired with the unique characteristics of the Bingham environment. The efficient chemical disintegration of the hill also depended on a level of precision mapping and planning that was still very new at the time. Recall that one of the key reasons General Electric and other financiers had been so leery of Jackling and Gemmell's initial proposal was that they did not trust their claims that the deposit was immense. Put simply, they did not believe the two men could really "see underground" nearly so well as they professed. This lack of trust occurred in part because Jackling and Gemmell were using a revolutionary and still unproven new technique of remote sampling through drill holes.[100]

In the early twentieth century, two new devices were available for remote underground sampling: churn drills and diamond drills. The first portable churn drill was built by an Iowa company in 1867 to drill water wells.[101] By the late nineteenth century, petroleum and mining engineers had adapted the drills for use in tapping oil fields and to explore subsurface geological structures. Jackling and Gemmell used a steam-powered churn drilling rig that rapidly lifted and dropped a heavy column of steel rods with a chisel-like cutting bit. Churn drills cut by percussion, as the bit strikes the rock at a rate of thirty to fifty times a minute, reducing the hard rock to powder. As the drill advanced, every five feet or so the operator flushed out the hole with pressurized water, which carried the fragments of rock back to the surface. The drill operators took careful samples of this "sludge" for further assaying.[102] Diamond drills developed at about the same time as the churn drills but were distinct in that they cut by the rotation of an O-shaped drill bit studded with small shards of "black" industrial diamonds. As the drill assemblage cut, a solid cylindrical core of rock was forced up into the hollow center of the drill bit and steel connecting rods. Periodically, the operator raised the drill and removed the cores while carefully noting the depth at which the sample had been taken. Later, geologists would study the cores to provide a highly detailed map of the rock strata at one point in the deposit.[103]

Whether using churn or diamond drills, sinking a long drill hole was expensive. At Bingham the cost of drilling a typical fifteen-hundred-foot hole was about thirty thousand dollars.[104] This was still far cheaper than the alternative: blasting out a six-by-six-foot tunnel of that length. Whether to use a churn drill or a diamond drill depended on the firmness of the rock and whether the geologist or mining engineer believed only drill cores could give a truly accurate sample of the rock. There was a lively debate over which method struck the best balance between cost and accuracy, as churn drilling was definitely cheaper. Diamond drills, though, had the advantage of providing solid cores; they could also drill on an angle or even upward into rock, which was impossible with the churn drill.[105]

At Bingham, Jackling and Gemmell used churn drills to probe the extent of the deposit in their initial 1899 survey. Though the technology had recently proven its accuracy in sampling the big Mesabi iron ore deposits in Minnesota, many still viewed it with some suspicion.[106] However, six years later when the Guggenheims debated whether to invest in the Bingham operation, their consulting engineer again tested the property using both diamond and churn drills. Only after the Guggenheims had the results of the drill hole tests did they agree to risk their millions.[107] Regardless of the method used, exploratory drilling gave engineers, financiers, and geologists much more power to make reasonably accurate estimates of the nature and extent of an ore deposit before any mining began.[108]

Closely related to the development of drilling technology was the growing importance of geological engineers and mining geologists.[109] These technical experts used the data from exploratory drilling to plan and execute what they hoped would prove the most efficient and economical means of sinking a pit. Serious errors in estimating the grade or extent of a deposit could prove disastrous to an operation where a slight change in the planned depth, size, or slope of a pit could increase or decrease the amount of overburden moved by millions of tons. Likewise, while disseminated ore deposits like Bingham were more uniform than a typical lode deposit like Butte's, the grade of the copper ore did vary. Nor was the best grade of ore to mine necessarily the richest ore, as in an underground mine. Open-pit geologists and engineers often mined and mixed the various grades of ore to maintain a uniform average best suited for concentration.[110]

Remote drilling and sampling techniques thus radically changed the design, financing, and operation of mining, and nowhere more so than in the new open-pit methods pioneered by Jackling. Whereas mining had once depended upon the skills of underground workers who identified and extracted only the best ore, Jackling's highly planned and rationalized mass destruction method transferred almost all such decisions into the hands of a few top managers, engineers, and geologists. Thanks to extensive drill sampling and geological mapping, the technical experts alone knew the location, depth, and extent of the various grades of ore in the deposit. They alone could "see" the future extent of the giant pit. Thus, while the people of Bingham may have only begun referring to "the Pit" in the early 1930s, Jackling and his engineers had been envisioning its shape and extent for years before that point. Far more than was the case with the typical underground mine of the previous century, the Bingham mine was an abstract idea before it became a concrete reality.

Prior to World War II, Jackling and his successors at Utah Copper sank 159 churn drill holes into the Bingham deposit, drilling a total of almost twenty-four miles.[111] Never before had such a large copper deposit been so thoroughly explored and mapped, and never before had humans been able to exert such complete and accurate control over an ore deposit's subsequent development. Paired with the precision power of high explosives, the geological mapping of the Bingham deposit allowed Jackling to plan and execute the efficient disintegration of a mountain of ore into a pile of rubble. But having succeeded brilliantly in streamlining the ore-breaking process, Jackling also needed a means of loading and transporting the many thousands of tons of broken rock every day. If drills, maps, and dynamite had given the engineers an Olympian power to envision the pit, they would also need a matching Herculean power to lift, load, and transport millions of pounds of broken ore.

STEAM SHOVELS: THE MOST HUMAN
OF POWER MACHINES

As the mining engineer and historian A. B. Parsons wrote in 1933, few individuals "have not watched with fascination the almost human motion, the more than human strength, the rhythm of the whole steam-shovel cycle of crowding, digging, swinging, dumping, swinging, and crowding again."[112] Americans had also been intrigued by earlier machines of the industrial era, like steam-powered railroad engines and boats. Yet the steam shovel was unique in its replication of an actual human motion: digging with a tool. To be sure, in their power of locomotion trains and boats shared in an even more fundamental property of most living creatures. These industrial artifacts, however, moved only by recourse to wheels, paddles, propellers, and other appendages for which there were no obvious human analogues. They rolled rather than walked, paddled rather than swam. In any event, even creatures like mice and fish could move, but so far as was known until recently, only humans used tools to dig. The steam shovel was striking, then, in its mechanical re-creation of the distinctively human action of digging with a tool, but with vastly more power and speed. Parsons was right: the motion of the steam shovel seems "almost human."

The steam shovel was the precursor to the modern robot and mechanical arm, machines that replicated or extended the natural biomechanical abilities of the human body. Especially after steam shovels were equipped with fully rotating platforms and caterpillar tracks, their remarkably humanlike abilities made them easily anthropomorphized. By far the most famous example of this transformation of a steam shovel into a gentle giant was the now classic 1939 children's book by Virginia Lee Burton,

Mike Mulligan and His Steam Shovel. Intriguingly, the author was the daughter of Alfred E. Burton, the longtime dean of students at Massachusetts Institute of Technology. Whether the technophilic MIT atmosphere influenced her choice of projects is difficult to determine, but she did seem to favor stories about humanlike machines. The star of her first book was a runaway little locomotive with the onomatopoeic name of Choo Choo.[113]

Burton's story and illustrations for *Mike Mulligan and His Steam Shovel* became her biggest success. Mike Mulligan is the proud operator of a friendly red steam shovel named Mary Anne, a play on the name of the American steam shovel manufacturing company Marion. Burton's illustration of Mary Anne is based on a typical small steam shovel of the 1920s with a caterpillar mount and fully revolving platform. Mary Anne, of course, also has eyes and a mouth on her dipper; the long shovel boom is her neck. The story itself is one of technological obsolescence. Despite years of faithful service in digging canals, railroad cuts, and skyscraper foundations, Mary Anne now faces the scrap yard as new electric and diesel shovels get all the jobs. To save his beloved shovel, Mulligan drives Mary Anne out into the countryside, where he promises the citizens of Popperville he can dig the cellar for their new town hall in just one day, boasting that "Mary Anne can dig as much in a day as hundred men can dig in a week." Mulligan and his shovel succeed, but in their haste fail to leave a ramp behind so that Mary Anne can crawl out of the new cellar. A bright young boy solves the problem when he suggests the steam shovel could serve as the furnace of the new town hall and Mulligan as the janitor. All agree, and in the final illustration Burton shows a proudly smiling Mary Anne, still complete with dipper and boom, while her coal-fired steam boiler is anchored to the floor and connected to steam heating ducts.[114]

Though Burton's steam shovel resembles a mechanical brontosaurus more than a human being, the book nonetheless offers a charming example of the American fascination with these machines. As Parsons asserted, steam shovels were the "outstanding symbol of the machine age," and in no small part because of their surprisingly humanlike movements.[115] Indeed, in the steam shovel's rough but potent replication of that fundamentally human action of digging lay its immense value in Jackling's system of mass destruction.

As the business historian Alfred Chandler demonstrates, the key to achieving mass production economies in industries from oil refining to metalworking was speed. Economies of size or scale were far less important than increasing the "velocity of throughput."[116] If a factory could use

new machinery and management techniques to make two hundred ciga-
rettes in the time it had previously produced fifty, sales could be quadru-
pled while the fixed costs for machinery, light, power, and maintenance
often changed little. In some light manufacturing industries, such as clock
making or woodworking, increased speed was achieved primarily through
adopting more efficient machinery or changes in plant design that minimized wasted effort. However, in his authoritative history of Ford's development of mass production, David Hounshell demonstrates that the
moving assembly line was the key innovation. By moving the work to the
men, Hounshell argues, "the Ford production engineers wrought true mass
production."[117] Likewise, in heavy industries like steel making, Chandler
notes that engineers adopted energy-intensive machines to move materials
through the factory far more quickly than human or animal power could.
As one historian notes of the production process at the Carnegie steel works
in Pennsylvania, "Steam and electric power replaced the lifting and carrying action of human muscle . . . and people disappeared from the mills."[118]

In Chandler's perceptive observation that speed was the essence of mass
production lay an equally profound insight into understanding modern
environmental degradation, though few have recognized it as such. Environmental historians initially paid little attention to his work, perhaps seeing many of his ideas as largely antithetical to their own. Chandler, who
died in 2007 at the age of eighty-eight, was, in the words of the *Economist*,
"the dean of American business historians."[119] Born into a wealthy Delaware family with close ties to the du Ponts (his middle name was du Pont),
Chandler attended Harvard, where one of his friends on the sailing team
was John Kennedy. He later returned to his alma mater as a professor in the
Harvard Business School and worked there for nearly twenty years until
his retirement in 1989.[120] Chandler was a prolific writer, but perhaps his
most famous and influential work was his 1977 book *The Visible Hand*,
which lays out the nature of mass production as part of his broader argument that the visible hand of modern corporate managers had largely replaced Adam Smith's invisible hand of the free market.

Though *The Visible Hand* won both the Pulitzer and Bancroft prizes,
neither the environmentalists of the time nor scholars in the emerging
field of environmental history paid much attention to the book. In part,
this neglect is understandable since Chandler did not even consider the
possible environmental consequences of the corporate "managerial revolution" he charts. Further, his argument for the technological and managerial logic of the modern corporation often seemed to justify big business

as both inevitable and inherently desirable. Since many environmentalists and environmental historians of the time viewed corporate capitalism as the major cause of environmental degradation, Chandler's seeming celebration of the architects of twentieth-century big business was unlikely to find much of an audience among them. There were, however, reasons to question a simplistic assertion that capitalism per se is the root cause of environmental degradation. Consider, for example, the abundant evidence of massive environmental damages in regions with state-run economies like those of the former Soviet Union.[121] This suggests the issue is less one of a specific political economy than of the ways in which elites of a variety of ideological stripes exploit both nature and other humans for their own benefit.[122] More to the point, if the capitalistic system itself was inherently destructive of the environment, then any effective solution must entail replacing it with some other political economy. There was little value, then, in pondering such detailed historical studies of capitalist business practices like those offered by Chandler, particularly if they seemed to suggest that some aspects of capitalism might also be of value. Thus the possibilities for any productive intellectual cross-fertilization between business history and environmental history seemed limited.

This was unfortunate, since an understanding of Chandler's ideas specifically and business history in general is critical to any sophisticated comprehension of modern environmental degradation. However, in 1992 another Bancroft Prize–winning book, William Cronon's *Nature's Metropolis*, began constructing an important and enduring bridge between the two fields. Drawing extensively on Chandler's earlier work, Cronon neatly ties the development of Chicago-based factories for the early mass production of meat, grain, and lumber with the resulting environmental consequences in the western hinterland where the raw materials originated. While skillfully laying out some of the environmental consequences of mass production industries, Cronon's book also echoes Chandler in its careful explanation of the economic and technological logic of large-scale high-throughput meatpacking plants and grain mills. To be sure, Cronon argues that the corporate managers of these enterprises were relentless in their drive to maximize profits, with often severe consequences for the natural environment. But in the process, he reminds us that they also provided cheap and abundant meat, flour, wood, and other essential products for a growing nation of consumers.[123]

In a sense, Chandler had been doing work important to environmental history all along, though it remained for others like Cronon to draw out the connections between his urban high-throughput factories and the surrounding natural world. Large-scale industrialized mining, however, does not really fit Cronon's model of urban factories driving the exploitation of natural resources in a distant hinterland. While western mine operators did typically ship their metals to urban manufacturing factories, it is critical to recognize that many of the western mines were also adopting the very same principles of high-speed throughput used in Chicago factories. The factories, in other words, were not just in Chicago or other eastern cities; they were in the rural hinterland itself.[124] Further, while Cronon's work effectively ties Chicago factories to the surrounding environment, in mining it makes no sense merely to link factory and environment: they were the same thing. In mining, the factory is embedded in the natural world, and any change in technological and managerial systems tends to have an immediate effect on ecological systems. In this sense, rural environments were not just exploited and often damaged in order to provide natural resources to distant urban factories. Rather, they were also more immediately and directly harmed by these hybrid environmental and technological factories, which were located in the countryside but followed many of the same principles as the urban factories.

In contrast to their urban cousins, achieving high-speed throughput in these rural envirotechnical factories entailed not a system of mass production but rather one of mass destruction. Again, Chandler's work suggests why. In a brief but insightful passage in *The Visible Hand*, Chandler observes that in mining "there was little opportunity to speed up the process of production by a more intense application of energy."[125] Chandler, of course, was referring to conventional underground mines where it was difficult to adopt big steam or electrically powered machines like those used in the Carnegie steel factories that had "replaced the lifting and carrying action of human muscle."[126] At the Carnegie plant, steam locomotives on thirty-inch-wide rails moved many raw materials, while powered rollers and big overhead cranes carried the heavy steel pieces through successive production stages.[127] The mine, however, was a very different sort of place. It might have been theoretically feasible to build a small steam locomotive underground, but the coal smoke it produced would have overwhelmed even the most powerful ventilation system. Likewise, it was one thing to build a system of powered rollers or conveyor belts in a steel plant where the floor was smooth and flat, quite another in the rough narrow passages of a mine where, to make things worse, the flow of ore was constantly changing as the mine expanded and followed new seams.

Most important, the slowest and most labor-intensive process in an underground mine was not in the horizontal movement of the ores. Mules and horses already provided considerable assistance in that work. Rather, it was in the mundane and backbreaking human task of simply lifting and loading the ore into mine cars with a wide-edged flat shovel. If the intensive application of hydrocarbon-powered machinery was to significantly speed up the mining process, it would need to be applied to this most basic of tasks, one that miners dismissively called "mucking" and relegated to the lowest and least experienced men.

In the early twentieth century, there were machines capable of making intensive use of energy to replicate and vastly speed up the human task of shoveling. But to have put the massive rail-mounted steam shovels to work underground would have been extraordinarily difficult, if not altogether impossible. If mining was to achieve high-velocity throughput, then, the design of the envirotechnical factory itself would have to be altered to permit the use of these big machines for shoveling. The subterrestrial factory mine would need to give way to the terrestrial factory pit. There were other advantages to open-pit mining beyond using steam shovels, of course. As discussed earlier, the underground mineshaft was a bottleneck that impeded the flow of materials in and out of the mine. Likewise, given the high costs of drilling tunnels, only a limited number of miners could simultaneously develop an ore vein underground. The open-pit mine eliminated both of these problems, doing away with the need for a shaft at the same time it laid bare acres of ore. An open-pit mine operator could, in theory, put as many men to work as could be safely crowded into the pit.[128]

Despite these other advantages of open pits, though, it is unlikely that Jackling's Bingham mine would have been nearly so profitable had it not been for the increased speed made possible by power shovels. Improvements in ore blasting, moving, and processing were all similarly beneficial to underground mining operations. Only the mechanization of the loading process with steam shovels made it economically feasible to remove the thick layers of overburden from ore deposits and rapidly load thousands of tons of rock blasted every day. Thus, more than any other technology, the steam shovel was critical to the success of Jackling's system of mass destruction. Thanks to the economies of speed made possible in large part by the power shovels, by 1944 open-pit mines had more than twice the average output per man-day than underground mines.[129] As one commentator rightly notes, "The history of open cut mining is largely the history of the power shovel."[130]

William Otis built the first practical steam shovel in Canton, Massachusetts, nearly seventy years before Jackling began mining. Otis mounted a dipper on a long boom anchored to the front of a modified railcar that allowed the boom and dipper to swing up and down and side to side, thus creating the basic power shovel design that would endure for decades to come. Otis put his new rail-mounted steam shovel to work excavating the route for the Boston and Providence Railroad, and it quickly proved a remarkable success.[131] After Otis died in 1839, his contracting partners continued to build and improve on the device. The partners eventually made three different shovel sizes, ranging from eight to sixteen horsepower, but they used the shovels exclusively for their own company and only built about twenty of the machines.[132]

Despite this slow start, a careful observer might well have recognized the economic potential of Otis's invention. Two historians who calculated the economics of the early Otis machines estimate that even these crude shovels could replace 60 to 120 workers, resulting in a twenty-year labor savings of about $225,000. Since an Otis shovel cost about $5,000 to $7,000, the investment would quickly pay off, provided the costs of fuel did not outstrip the savings in labor. Given these numbers, when certain key patents on the Otis shovel expired in the 1860s, machinery makers around the nation began to construct and market shovels based on the Otis designs.[133] The machines changed little over the next few decades until specialized steam shovel manufacturers began to appear between 1880 and 1890, including the two major players, the Marion Steam Shovel Company and the Bucyrus (later Bucyrus-Erie) Company.[134]

Railroad contractors bought almost all the power shovels made in the United States until the late nineteenth century, as the machines proved indispensable in cutting rail lines through hills and leveling grades.[135] Gradually, other industries began to recognize the potential value of the shovels, and Marion and Bucyrus responded by building a wider array of specialized excavators. In 1892, Chicago contractors bought twenty-seven Bucyrus steam shovels to dig a new city irrigation drainage canal, marking the first time excavators employed a large battery of power shovels for heavy digging. The success of the Chicago canal project sparked increased use of shovels for large-scale construction projects, including the widely publicized and celebrated excavation of the Panama Canal, which used seventy-seven Bucyrus-Erie shovels and twenty-four Marion shovels to move over 255 million cubic yards of material.[136]

Despite the fact that mining is often a matter of moving large amounts of dirt and rock, mining engineers and others in the industry initially paid little attention to the shovels. As already noted, the huge coal-fired machines were obviously impractical for use in the underground mines that dominated in the hard rock industry. The strip mining of coal, however, was different. The first recorded instance of coal strip mining occurred in 1866 in Danville, Illinois, where an operator used wheelbarrows and horse-drawn scrapers to peel the earth from a shallow coal seam. A decade later, two contractors who had learned to use an Otis steam shovel on a railroad construction project adapted a similar machine for stripping a coalfield.[137]

Using steam shovels to strip-mine coal made sense. Many eastern coal deposits were in the form of large, easily broken coal seams covered by a thin layer of dirt and gravel. The same was also true for the Mesabi iron ore deposits in Minnesota, where an operator first began using a steam shovel in 1893.[138] On the Mesabi, much of the ore was already exposed to the surface or, at most, covered with only fifty feet of dirt and gravel. Under these conditions the costs of steam shovel surface mining was only fifteen cents per ton, a third of the typical forty-five cents per ton required for underground operations.[139]

Historians have long debated who first suggested the use of steam shovels to mine the Bingham low-grade copper deposit. Some have credited Samuel Newhouse, the operator of a Bingham Hill mine just above Jackling's, or a consulting engineer who worked with De Lamar along with Jackling.[140] But as Parsons rightly notes, the 1899 Jackling-Gemmell report was "the first conservative and reasonably comprehensive analysis of a mining enterprise based on the exploitation of ore containing as little as two per cent copper, or forty pounds to the ton."[141] Perhaps many mining men considered using open pits to mine low-grade copper ore, as the success of the shovels and pits on the Mesabi iron range was widely publicized. But it remained for Jackling and Gemmell to make the first informed and serious proposal outlining exactly how such an operation might be profitably achieved. Further, it is often the case in the history of innovation that the challenge is not so much in coming up with the basic idea but rather in making it into a working reality. In this sense, there is little question that Jackling's work was paramount.

In realizing his plan, Jackling was also fortunate in having the assistance of his original partner in writing the 1899 proposal for De Lamar. In January 1906, Jackling lured Robert Gemmell away from his work in Mexico and appointed his old friend as general superintendent of Utah Copper.[142] Now both men had an opportunity to prove that their ideas, which the mining community had so roundly rejected seven years before, could really work. They acted quickly. Shortly after joining Utah Copper, Gemmell accompanied his new boss on an inspection trip to what Jackling referred to as "the most important steam-shovel operations in the United States, more particularly in the [Minnesota] iron regions."[143] Following this inspection tour, Jackling hired an experienced Minnesota open-pit manager named J. D. Shilling as mine superintendent.[144] Under Shilling's guidance, three rail-mounted steam shovels were stripping the Bingham hill by that August, two of them built by Marion and the third by Vulcan (a small manufacturer later absorbed by Marion). The shovels commenced the huge task of removing the covering of shrubs, wild grasses, earth, gravel, and oxidized hard rock that overlay the sulfide copper ore beneath. Stripping at a rate of about a hundred thousand tons per month—roughly an acre-sized chunk of hundred-foot-thick overburden—by June of the following year the shovels had exposed nearly six acres of ore.[145]

"The digging unit is the 'heart' of an open-pit operation," as two mining engineers rightly noted some years later. "Other phases are carried on to keep the shovels busy."[146] To use the fuel-hungry power shovels at Bingham profitably, Jackling's drilling and dynamiting crews had to ensure a constant supply of properly broken ore. In order to keep the steam boilers warm, the shovels had to burn coal even if they were sitting still. If the ore breaking ever lagged behind so that the shovels were idle for long, the cost advantages of power loading quickly went up in coal smoke.[147] The development of improved explosives and drilling made this rapid ore-breaking pace feasible. Black powder would have been too slow and imprecise. But in the next twenty years, the Bingham dynamite crews steadily increased the pace and size of their blasts just to keep up with the rapidly improving ability of the power shovels to move more material at a faster pace.

One of the first major technological improvements that Jackling adopted at Bingham was the fully revolving shovel.[148] With the old Otis-type shovel, the boom was attached to the front of a conventional railroad car. As a result, the operator could only move the shovel about 180 degrees from side to side. With the fully revolving shovel, the entire shovel platform rotated on its carriage—housing, engine, operator, and all—allowing the boom to move in a full circle. This 360-degree range of motion greatly increased the power shovel's digging flexibility and speed. When digging, the crew blocked the wheels in place to keep the shovel from sliding back on the rails. Some shovels also needed to set extendable braces to keep them from tipping over on their side under the weight of lifted ore.

12. One of Jackling's early rail-mounted steam shovels stripping overburden in 1912. The success of Jackling's system of mass destruction mining depended on the substitution of hydrocarbon power for human power with these crude coal-fired steam shovels. Though slow and cumbersome in comparison to later machines, the shovels could still move more ore in one or two minutes than a strong man could move in an eight-hour day. *Used by permission, Utah State Historical Society, all rights reserved.*

Moving the shovel was thus a time-consuming process. A fully revolving shovel was substantially faster because the operator could remove a wider swath of ore before having to move the shovel. An even more important time- and labor-saving feature was that the shovel could now pick up its own railroad tracks from behind and move them to the front as it traveled along. This feature eliminated much of the slow heavy lifting and carrying previously done by a sizeable track-laying crew.[149]

In yet another case of the many feedback cycles between copper and electricity, Jackling also eventually adopted electrically powered shovels. The electric shovels offered an attractive alternative to the expense and mess of steam engines. Every steam shovel engine in the pit needed constant feeding via a two-inch water line. To satisfy the shovels' appetite for coal, dedicated supply trains scurried back and forth over the narrow mine benches, interfering with the progress of blasting and mining. Maintenance of the steam shovels was also laborious. Workers washed out the boilers every day, and the steam engines kept a corps of mechanics busy repairing frequent problems.[150]

An iron-mining company on the Mesabi was the first to use an electric shovel, purchasing a custom-made machine from the General Electric Company in 1920.[151] As a major investor and promoter of electric power in Utah, Jackling had more than one reason to be an early proponent of electrification of mining, and he bought the first two electric shovels for Bingham in 1923. He continued to use several steam shovels well into the 1930s, though not because he shared Mike Mulligan's nostalgic affection for the outdated machines. Since Jackling had a cheap and ready supply of coal available from the company's nearby mines, the coal-fired steam shovels continued to be reasonably efficient despite their many shortcomings.[152] In the long run, though, the many advantages of electric shovels outweighed the cost savings offered by even the cheapest coal. Electric shovels cut labor costs by eliminating the need for a fireman and increased the pace of mining by ending interference from coal supply trains. The electric shovels were especially attractive during Bingham's often subfreezing winter days, when the water supply lines for the old steam shovels had tended to freeze. As a result, operating costs for the electric machines were about 23 percent less than for steam, and labor costs were 80 percent less. As an added bonus, the electric shovels cut down on accidents. In the era of the steam shovels, the puffing mechanical monsters could often dangerously obscure the track and banks in dense clouds of steam and smoke. Even in coal strip-mining operations, most companies found it was cheaper to use electric shovels rather than feed their own coal to steam shovels.[153]

The final major improvement in the Bingham power shovels was also the most obviously intriguing to the typical observer, then and now: the introduction of caterpillar traction mounts. The Bucyrus-Erie Company began offering a broad-faced, slow-moving traction wheel mounting as an option on its shovels as early as 1912, but the wide wheels performed poorly in the rough pits. A truly viable alternative to railroad tracks demanded a new application of the very old idea of caterpillar tracks.[154] An English inventor patented the basic principle of caterpillar traction in 1770, but the materials of the day were too weak to construct a workable device.[155] The credit for the invention of a workable caterpillar technology went instead to a California farming machinery designer named Benjamin Holt. Holt worked near a reclaimed marshy area outside of Sacramento, and farmers frequently complained to him that their heavy tractors often sank into the soft earth and became stuck. To solve the problem, Holt turned to the simple idea of dividing a basic round wheel into hinged segments extending over the length of the vehicle: in essence, a device in which the vehicle laid down and then picked up its own track as it went along. Holt successfully applied his invention to tractors in 1904, calling it the "caterpillar." Within a decade, farmers were using more than two thousand Holt caterpillar-equipped tractors in twenty countries around the world. The utility of the device for crossing rough terrain appealed to other inventors as well. During World War I, a British army engineer adopted the technology for more nefarious purposes by using it on early tanks.[156]

The caterpillar-mounted power shovel eliminated the need for the four- or five-man track-laying crews that had previously serviced each rail-mounted shovel.[157] For this huge savings in labor costs alone, the mining engineer E. D. Gardner considered caterpillar mountings to be the most important power shovel innovation of the twentieth century. The greater maneuverability of caterpillar shovels also sharply reduced the risk that shovels and their crews would be buried under an avalanche of falling ore. Because with the old railroad mounts pit crews had to block the wheels of the shovel into position while digging, it took the crew at least ten minutes to prepare a railroad shovel to back up. By contrast, caterpillar shovels could immediately back away if there was any danger of a rock avalanche and then quickly return to resume digging. Again, the speed of lifting the ore from the ground to the waiting railcar increased as a result.[158]

By the time America had entered into World War I, Jackling had settled on the basic steam shovel design best suited to the rapid movement of ore and waste rock at Bingham: a fully revolving electric shovel on a caterpillar mount. He had begun his open-pit mine in 1906 with three modest shovels. Twelve years later, twenty-one shovels were digging away at the hill, most of which were Marion Model 91s. The typical Bingham steam shovel had a dipper capacity of about four cubic yards, which meant the shovel could lift a cube of rock about one and a half yards on each edge. This is roughly the volume of a modern garbage Dumpster, though at about sixteen thousand pounds, this amount of Bingham rock weighed a bit more than a typical Dumpster full of cardboard and spoiled produce. In a 1919 paper published by the American Institute of Mining Engineers, the "Efficiency Engineer" for the Phelps Dodge Corporation took to heart the recommendations of Frederick Taylor and other advocates of so-called scientific management and conducted a careful study of human shoveling. The efficiency engineer was disappointed that his "poor grade of Mexican" laborers managed to shovel only about nine tons of ore in a shift. According to his measurements, better techniques should permit them to shovel at least eighteen tons in eight hours of work. Leaving aside the author's racial slurs and suspect "scientific" methods, the paper nonetheless is probably correct that, pushed to an unsustainable extreme, a healthy man might at best shovel eighteen tons of ore into a four-foot-high ore car in a shift.[159] The power of the steam shovel is obvious by comparison. Even the relatively small early shovel used at Bingham could move that amount of ore in mere minutes with two or three scoops of its four-cubic yard dipper. In those stark numbers lay the odd logic of mass destruction mining. Only the open pit allowed Jackling to use these machines and thus achieve the economies of speed of a Carnegie steel mill.

Given the immense increases in speed they offered, the size of the shovels used at Bingham grew steadily. By the 1950s, Bingham had several Bucyrus-Erie 190-B shovels with eight-cubic-yard dippers.[160] With a capacity twice that of the shovels used in the 1920s, the 190-B could have scooped up and lifted a small automobile as easily as a child playing with a toy car. Even these big machines were tiny in comparison to those used in more recent years. Today, Bingham's P&H electric shovels are the size of a ten-story building and have thirty-four-cubic-yard dippers. Fifty people could stand inside one of these dippers without being overly crowded.[161] Even bigger shovels are available, though Bingham's relatively narrow benches and the need for frequent movement would likely make them inefficient. In strip mining on flat ground, the trend toward gigantism has been even more pronounced. In 1965, Marion built what continues to be the world's largest caterpillar-mounted shovel. Nicknamed "the Captain," the shovel was the biggest mobile land machine ever built. Designed to strip the earth from shallow coal seams in Illinois, the shovel's dipper had a capacity of 180 cubic yards, roughly the dimensions of a small six-hundred-square-foot house. The shovel itself was twenty-one stories tall and weighed twenty-eight million pounds.[162] Even caterpillar tracks could not keep such a behemoth from sinking into the ground, and operators had to place three-foot-thick wooden mats under its tracks as it slowly crawled forward. After almost thirty years spent stripping immense amounts of land, one house-sized bite at a time, a fire in the shovel's hydraulic system finally brought the Captain to a halt. The owners sold it for scrap in 1992.[163]

The steady growth in size of power shovels and the improvements in their flexibility, power, and traction all had one goal: to increase the speed with which they moved material, whether it was Illinois soil or Utah copper. Even a tiny four-cubic-yard shovel could move rock more quickly than a hundred humans, and the big shovels could do the work of thousands in mere minutes. They did so, of course, only by using immense amounts of energy. Jackling and the many subsequent open-pit mine operators had found a way to achieve the "intense application of energy" that Chandler argues eluded underground mining operations.[164] In 1911, the engineers at the Bucyrus Company did some intriguing calculations. Based on their statistics, they concluded that that a small two-cubic-yard steam shovel consumed about 3 tons of coal in order to load 192 tons of ore. This steam shovel would thus consume about 3,125 tons of coal in lifting and loading 200,000 tons of Bingham's 2 percent copper per month, which in turn would yield about 4,000 tons of copper.[165]

In sum, for every pound of copper the Bingham mine produced, Jackling had to burn about three-quarters of a pound of coal simply to lift the ore from the floor of the pit and dump it into a railcar. Or, to return to a particularly symbolic piece of copper, if Thomas Edison's cubic foot of copper had come from the Bingham hill, a steam shovel would have consumed 365 pounds of coal in its creation. This energy calculus does not even include the coal and other energy sources consumed to transport, concentrate, smelt, and refine the ore. Larger shovels powered by electricity, or later diesel, improved the energy efficiency, but they did not alter a fundamental reality: open-pit copper mines only made economic sense because they drew deeply on the world's finite supply of energy resources in order to increase the speed of mining. More than any other single technology, the energy-hungry power shovel had made Jackling's high-speed and high-throughput copper mining system feasible. Indeed, it is not too much to say that the steam shovel was nothing more than a device to permit Jackling to channel large amounts of concentrated hydrocarbon or hydropower energy into the previously slow and labor-intensive process of mining. The odd logic of mass destruction thus stemmed in no small part from the desire to use these crude but powerful machines.

Precisely because steam shovel mining was so crude, however, the final stage of the mass destruction process had to be anything but. The very imprecision and speed of the immense dynamite explosions and giant shovels meant that the process of removing the tiny amounts of copper from vast amounts of ore would have to be almost insanely precise.

CONCENTRATORS: THE ART OF EXTRACTING
A NEEDLE FROM A HAYSTACK

Imagine you are given a wheelbarrow filled with one hundred pounds of what appears to be very fine-grained sand, nearly the consistency of bread flour. Although you cannot see them, scattered among the many millions of grains of worthless sand are a much smaller number of pure gold, perhaps one grain of gold for every hundred of sand. Since the entire load weighs about a hundred pounds, you quickly estimate that the wheelbarrow must contain at least a pound of pure gold, an amount that would be worth about ten thousand dollars. If you can somehow separate the gold particles from the sand, this ten-thousand-dollar golden treasure is yours. If not, you have nothing more than a wheelbarrow of worthless fancy sand that is too fine even to be used in a children's sandbox.[166]

The question, of course, is: how do you separate the gold from the sand?

While a good deal more complex and involving copper instead of gold, this was essentially the question that faced Daniel Jackling in the early years of mining at Bingham. It was no secret that an immense deposit of low-grade copper ore lay hidden under the Bingham hill. But even the biggest deposit of copper, or any other metal, was nothing more than worthless rock if the technology did not exist to separate the coveted minerals from the mass of worthless rock. Engineers call this mineralogical version of separating the wheat from the chaff the process of ore concentration. One of the earliest and simplest concentrating devices was the classic prospector's gold pan of forty-niner fame. Prospecting pans were broad, shallow dishes in which the prospector would swirl a small amount of promising alluvial soil with water. When this was done properly, the moving water washed away the lighter sand and dirt and left behind the heavier particles of gold. Many more complex and mechanized separation and concentration technologies used this same basic principle, though with hard-rock mining the ore had to first be crushed in order to free the particles of metal. All such concentrating techniques depend on the fact that gold, copper, and other metallic minerals have a greater density or specific gravity than sand, soil, and most other minerals. Hence mining and metallurgical engineers referred to them as gravity concentration methods.

At first blush, gravity concentration might seem just the thing to winnow out the pound of gold from your wheelbarrow of sand. But in actual practice, generations of mine operators had found that successful gravity concentration was often maddeningly difficult, particularly when the material being processed was finely pulverized. Given the bread-flour-like consistency of the wheelbarrow of gold and sand, many gravity concentration methods would lose more of your gold than they saved. Even with material of a much coarser consistency, nineteenth-century operators found that their gravity concentration machines lost as much as 20 to 30 percent of the gold, silver, lead, and copper in their ore to waste. Indeed, so much gold, not to mention other minerals, was lost that mine operators knew that their own tailings (waste piles) would be worth a fortune if only someone developed an effective method to separate the minerals from the waste. Many ambitious inventors tried their hands at the challenge, but for decades a major breakthrough eluded them all. Instead, engineers and mine operators pushed the capabilities of gravity concentration to their limits of refinement and precision. By the early twentieth century, the typical concentrating mill contained a complex maze of grinders, regrinders, classifiers, and gravity concentrating machines. Engineers made detailed flow charts to better understand and refine the process, and they recycled materials through the system repeatedly in an attempt to squeeze out just a few more particles of metal.

The greatest enemy to successful gravity concentration was something mill engineers gave the appropriately descriptive name of "slime." In the process of crushing or grinding up the hard-rock ore so that it could be concentrated, inevitably some of the ore became much more finely pulverized than desired. When these "fines" entered the concentrator circuits, the tiny particles of heavier metal and lighter rock both became suspended in the water, rather like dust you might see floating in a sunbeam of light. This suspension of the particle largely negated the differences in specific gravity that the concentrating machinery depended upon to achieve separation. If you tried to concentrate your wheelbarrow of finely ground gold and sand using such early methods, slimes would have carried away most of your gold. On a much bigger scale, concentrating operations of all types lost millions of dollars worth of metals every year to "sliming."

The problem of slime was especially vexing with Jackling's low-grade copper ores at Bingham. Recall that Bingham's porphyry ore consisted of tiny specks of copper ore finely disseminated in a mass of valueless rock. To liberate these small particles of copper ore from the waste rock, Jackling had no choice but to crush the ore to a fairly fine degree, although not quite yet to the consistency of bread flour. At his Magna concentrating mill, Jackling continued the process of disintegrating the Bingham Hill begun by the dynamite blasts. After the steam shovels loaded the broken rock into ore cars, steam engines transported the material to the Magna mill in trains forty cars long. The ore first went through the big gyratory crushers that could break up chunks as big as four feet across. A series of Chilean rolling mills continued the process by smashing the rock between pairs of six-foot-high horizontal hardened steel drums.[167] The noise made by all these massive steam-powered rock crushers was deafening. The big problem with the crushing operations, though, was that they unintentionally but inevitably produced fines, which in turn became slimes during the concentration process. As a result, in the early years at Bingham, sliming carried away a punishing 30 to 40 percent of the profitable copper and other metals.[168]

Other engineers and inventors had previously tried to beat the challenge of porphyry copper ore slimes, but with only limited success. During the 1880s, a British-trained mining engineer named James Colquhoun pioneered the use of new machines called "vanners" to concentrate fairly low-grade copper ores at the Clifton-Morenci district in Arizona. Vanners used a continuous shaking rubber belt to move ground ore up a slight incline while a current of water flowed downward, washing back the lighter particles of waste rock so that only the heavier metallic ore made it to the end of the belt and fell into hoppers. Vanners did not solve the slime problem, yet they succeeded in capturing a greater percentage of the very fine mineral particles than had previous methods.[169] Thanks to the vanners and careful experimentation, by 1897 Colquhoun's concentrator was able to eke out enough copper to make a small profit on 5 percent porphyry ore. This, in Colquhoun's own view, "was the first porphyry concentrator designed and brought to success, [and it] meant that ores which up to that time had been thrown into the waste dump as worthless, had become of priceless value."[170]

At just about the same time that Colquhoun was improving the concentration of copper ores, a Colorado mine and mill operator named Arthur Redman Wilfley was also trying to do the same with low-grade silver and gold. Like Jackling, Wilfley was a Missouri native who moved to Colorado to try his hand in mining. He worked in an underground mine in Leadville, but he eventually managed to acquire his own silver mine in the city. Although he never received any formal technical training, he was a dedicated reader of mining books and journals.[171] Wilfley became obsessed by the thought of all the gold and silver sitting in the big tailings dumps scattered all around Leadville. Most of these tailings, and indeed

13. A big gyratory rock crusher destined for use in Jackling's concentrating mills. Chunks of copper ore as big as four feet across were fed into the mouth of this crusher, beginning the process of reducing the hard rock down to the consistency of bread flour. Note the worker standing alongside the exit chute for scale. *Courtesy Wisconsin Historical Society, Image (X3) 37408.*

many ore deposits, actually contained varying amounts of several different ores of copper, gold, silver, and other minerals. But since these minerals are all fairly dense and heavy, traditional gravity concentration methods and even the new vanners could, at best, only separate them all from the waste rock, not from each other. In 1892, Wilfley began a systematic effort to develop an improved method of concentrating for just such complex mixes of very low-grade materials. After at long series of experiments, Wilfley eventually came up with a new kind of concentrating machine: a broad, flat horizontal platform with a series of moveable parallel riffles of unequal length that terminated on one end in an oblique line. With careful adjustment, an operator could position each riffle to take advantage of very small differences in specific gravity, allowing the table to isolate and capture the various minerals in an ore.[172]

Wilfley began manufacturing and marketing his invention in 1896 and enjoyed immediate success. He soon signed an agreement with the giant Denver-based Mine & Smelter Supply Company to market his tables.[173] The company hired Wilfley as a staff engineer and put him to work setting up his machines in concentrating mills across the West.[174] Initially, gold and silver miners purchased most of the Wilfley Tables, putting them to use in "mining" their own tailings piles as much as for newly extracted ore. But in 1899, Wilfley and the Mine & Smelter Supply Company attempted to break into the burgeoning copper concentrating industry. The company first offered free demonstrations of the Wilfley Table in the rapidly declining Michigan copper mines, showing that it provided good returns and used less water than conventional concentrating machines. Success in Michigan soon opened the doors to Montana, where the Anaconda Company purchased dozens of the tables for its concentrator at the giant new Washoe milling and smelting complex.[175] Thanks in part to the new demands from the copper industry, by 1922 the company had sold more than twenty-six thousand Wilfley Tables.[176]

When Jackling designed his Magna concentrating mill, he was well aware of both James Colquhoun's work and Wilfley's impressive new table. He also knew that should his concentrating mill fail, all the other parts of the big open-pit mining system he and Gemmell had envisioned would also fail.[177] By using the giant steam shovels in an open pit, Jackling was violating one of the cardinal rules of good mining practice: extract only the valuable ore, leave behind the worthless rock. For millennia, this is what skilled miners had done. They followed veins or sheets of ore underground and minimized the laborious work of extracting, loading, and carrying it back to the surface by taking only what was of value. Indeed, a good working definition of the art of traditional underground mining might be "to find and selectively remove ore." But if Jackling was going to take advantage of the economies of speed made possible by steam shovels and open pits, he would have to engage in "nonselective mining," though such a term is nearly an oxymoron by the standards of an earlier age. What Jackling proposed to do to the Bingham hill was not so much mine it as to disintegrate it.

Though earlier miners at Bingham had used underground methods to select only the richer ore, Jackling knew such an approach would be too costly in mining the much bigger deposit of low-grade disseminated ore. He hoped high-speed blasting and the powerful but imprecise steam shovels would make it economically feasible to abandon such slow selective mining techniques and essentially remove the entire mountain of ore.[178] Appropriately for a man trained as a metallurgical engineer, Jackling would assign the real task of selective "mining" to his concentrator. When Jackling's Magna concentrator began operations in 1907, it used many of the most advanced techniques available, including fifty-two of the vanners that had worked so well for Colquhoun and forty-eight of Wilfley's tables. With repeated cycling of materials between these machines and others, Jackling managed to recover just under 70 percent of the copper ore.[179] After another infusion of Guggenheim capital, Jackling doubled the size of the Magna concentrating mill in 1908 to six thousand tons per day. Water for the thirsty concentrating process came from nearby springs and the Jordan River. During periods of low flow, the company also drained and reused water from its own tailings ponds, where it stored the immense amounts of watery waste pouring from the mill. Jackling and his engineers had designed the original tailings ponds at Magna to be 1,315 acres, or about two square miles. But since 96 percent of the "ore" (though only about 30 percent of the copper) ended up in these ponds, they soon proved to be too small. By 1918, Jackling had expanded the ponds to about six thousand acres and begun steadily covering these eight square miles of land with a thick layer of fine tailings.[180]

Slime, however, continued to be a serious problem. Despite Jackling's best efforts, the Magna concentrator still sent 30 to 40 percent of the Bingham copper to the tailings ponds. Despite these losses, the huge cost savings from the open-pit mining method allowed Utah Copper to make a good profit. If a solution could be found to the problem of sliming, however, the resulting metals would be a direct addition to the mine's income stream. In fact, the solution to Jackling's copper sliming problem had already been slowly developing for more than two decades by that point. In 1885, a self-taught chemist and metallurgist named Carrie Everson first patented a crude method of separating metals from waste by using the tendency of oils to float on water. She had noticed that when a very finely ground mixture of rock and ore was immersed in a vat of oily water, the oil tended to cling only to the metallic particles, allowing some of them to float on the surface while the particles of waste rock sank. This basic idea—that the selective attraction of oils to metals could be used for separation—was an important part of the process that eventually became known as flotation. However, it was only one part. No workable system was possible based on the Everson patent. A British inventor patented a similar flotation process in 1898, and although he actually attempted to market his technique, it had little more commercial potential than Everson's system.[181]

The basic problem with all of the early techniques for flotation was that their inventors had not realized the importance of air bubbles. The upward lift provided solely by the buoyancy of light oil floating in water was too small to raise very much ore to the surface. Rather, the secret to successful flotation lay with a secondary phenomenon: the tendency of small air bubbles to cling to the oil-coated metallic surfaces and loft the ore particles upward like tiny balloonists. It would be several more years before an Italian inventor came up with a system that actually added air to the oily mixture with frothing paddles, infusing the mix with bubbles. A British company called Minerals Separation Ltd. bought the rights to use the Italian air-frothing system. After performing extensive experiments at a mill in London, the company finally arrived at the modern froth-agitation method of flotation, which achieves separation by the adherence of air bubbles to metallic particles coated with small amounts of oil.[182]

Flotation thus did not emerge suddenly and fully formed from the mind of one brilliant inventor. Nor was flotation a one-size-fits-all technology; rather, it had to be painstakingly adapted to each type of ore and mill. As the mining engineer T. A. Rickard noted in 1917, "It is to manipulation, learned empirically in the laboratory and mill, that the flotation process owes its metallurgical success."[183] Because of intense patent litigation and the need for extensive experimentation to build a workable system, interest in flotation only grew gradually in the United States, despite its revolutionary potential.[184] Jackling first observed flotation in action in Butte, Montana, where he was offering his hard-won expertise from the Magna mill as a highly paid consultant to the Butte & Superior Mining Company in 1912.

Impressed by its potential, the following year Jackling directed one of his metallurgists, T. A. Janney, to begin experimenting with flotation on the Bingham ore. During the next few years, Janney designed his own copper flotation machine, a device that precisely agitated the oil and ore mixture to obtain the ideal degree of frothing. Janney also determined that the addition of small amounts of mineral, pine, and creosote oils yielded the best results with Bingham ore.[185]

14. Hundreds of flotation machines at the Magna concentration mill. The frothy mixture of water, air, and finely ground ore caused the valuable copper and other minerals to float to the surface for easy removal. By increasing the copper recovery rate by more than a third, flotation made it possible to mine extremely low-grade ores. *Library of Congress, Prints and Photographs Division, Historic American Engineering Record, Bingham Canyon Mine, Reproduction Number HAER UT-5.*

Mass Destruction

By 1917, Janney's work had demonstrated that the Utah Copper Company could achieve substantial savings by adopting flotation, and over the next four years Jackling rebuilt both the Magna mill and his Arthur mill (acquired in 1910 from the Boston Consolidated Company) to incorporate Janney's machines. The results proved well worth the investment. Between 1905 and 1917, the two mills had recovered a combined average of only 61 percent of the copper using conventional gravity separation, losing most of the rest to slimes. With the addition of flotation circuits to treat the slime, the recovery rate rose to 81 percent by 1923 and would eventually reach 90 percent.[186] Ironically, flotation actually worked best with the troublesome slimes that had plagued Jackling since he first began mining and crushing the low-grade Bingham ore. Increasingly, Jackling and others would deliberately crush all their ore to a flourlike consistency and separate out the metals using immense ranks of frothing flotation machines.

Flotation, then, would be a very effective way of getting that pound of gold out of your wheelbarrow of sand. The powdery consistency of the material that had previously been a problem is now an advantage—no need to first do the very fine crushing required for flotation to work.[187] Just as flotation could transform your worthless sand into ten thousand dollars' worth of gold, so did it transform thousands of previously worthless low-grade copper deposits around the world into valuable mines. But what if instead of one gold particle in a hundred, your wheelbarrow contained only one in a thousand? Would it still be valuable ore? What about one in ten thousand or even one in a hundred thousand? In fact, by the late twentieth century, gold mining companies would use mass destruction methods to routinely mine "ore" that was only 7/10,000 of 1 percent gold. The effect was less dramatic with copper because of its far lower market price, but the definition of what constitutes copper ore had been similarly expansive. When Jackling first contemplated mining at Bingham in the 1890s, a deposit had to have at least 3 to 4 percent copper to constitute "ore." Thanks to his low-cost mass destruction technique, Jackling soon proved he could make a profit mining ores with less than 1 percent copper. In the late 1920s, Jackling's Utah Copper Company even began to boast that it no longer had *any* worthless waste rock in the mine, only "low and lower grade ore."[188] In recent years, the average ore grade of American copper mines has further declined to less than one-half of one percent.[189] The only limit on how much lower that percentage might go seemed to be the price—both economic and environmental—that human beings were willing to pay for more copper.

In 1940, Daniel Jackling penned an article for the *Journal of the Western Society of Engineers*, which he titled "The Engineer's Province and Obligation in Organized Society." The article probably came as a bit of surprise to the journal readers, as Jackling had never been a prolific writer and commentator like Raymond, Rickard, or Joralemon. Jackling was seventy years old then and long since retired from Utah Copper and active mine management. Still, he had spent more than half a century working in and observing the American and world mining industry. Perhaps he felt the need to share his hard-won insights. Whatever his motives, Jackling chose to address not the role of the mining or metallurgical engineer but rather that of "the engineer" broadly defined. Jackling had no doubt about the importance of the profession. The engineer, he writes, is "responsible, primarily, basically, and almost wholly, for mankind's welfare and the progress of civilized advancement in practically all phases of human endeavor and association."[190] Given this weighty responsibility, Jackling concludes that the engineer must find technical solutions to all obstacles, "because the economic and sociologic destiny of mankind must rely upon his achievements and guidance more than upon all other classes and orders of human relationships combined."[191]

Jackling no doubt considered himself to have been just such a skilled and benevolent engineer of the "destiny of mankind." His peers had awarded him nearly every honor the profession had to bestow, and he was widely seen as among the greatest mining engineers of the century. Thirty years earlier, Jackling was even presented with the chance to guide human destiny in a different way when the Guggenheim and Standard Oil interests offered to back him for a U.S. Senate seat in 1909.[192] Jackling declined. Perhaps even then he considered politicians and "other classes" of societal leaders to be of small importance in comparison to engineers like himself. By 1940, his Bingham mine and its many imitators had essentially guaranteed the nation would enjoy cheap copper for decades to come, an accomplishment with indisputably profound social and economic consequences. If anything, the copper mining industry had been too successful. During the Great Depression, overproduction and slack demand sent copper prices as low as five cents a pound. All but the most efficient copper mines fell silent.[193] As Jackling was writing his article, though, demand was again on the upswing. Soon the broadening war in Europe and in the Pacific would begin consuming immense amounts of copper in pursuit of another type of mass destruction. Before he died in 1956, Jackling would also witness the seemingly insatiable postwar consumer desire for homes, refrigerators, and televisions, all of them heavily dependent on copper. Prices would reach record highs of forty cents a pound, and experts would again raise fears of global shortages.[194]

Whether to win a world war or to feed mass consumption, Jackling and his colleagues had found a means to provide the copper. It was unquestionably "an achievement of engineers," as Parsons wrote.[195] In the next chapter, the central role played by politicians and "other classes" in encouraging the growth of a copper-hungry society will become evident. Jackling and other engineers were only feeding a demand whose root causes they understood poorly, if at all. Likewise, while Jackling and other engineers liked to portray themselves as noble public servants, many of them also became fabulously rich during their years of "service." Whatever his psychic rewards for guiding the fate of modern civilization, it surely did not hurt that Jackling had also become a millionaire by the time he was forty. In addition to his sprawling Woodside, California, mansion (now famous only because of Steven Jobs's attempts to tear it down), Jackling purchased the entire top floor of San Francisco's finest luxury hotel as a second home. To ease both his business and pleasure travels, he commissioned a half-million-dollar 267-foot yacht. Just ten yards short of the length of a football field, it was for a time the largest such vessel on the Pacific Ocean. Jackling named his big yacht the *Cyprus*, after the Mediterranean island where the ancient civilizations of the region first mined copper. The *Cyprus* had ten bedrooms, each with a private bath, a music room with an open fireplace, a library paneled in Tibetan mahogany, and, of course, copper and brass fixtures throughout. With its typical crew of fifty, the boat cost about twenty-five thousand dollars a day to operate.[196]

Wealth had always been Jackling's goal, and now the poor orphan boy from Missouri had become very wealthy indeed. So had the Guggenheims, the Rockefellers, and Jackling's other backers. It would be a mistake, however, to conclude from this that Jackling was therefore insincere or merely self-serving when asserting that engineers like himself were chiefly responsible for "mankind's welfare and the progress of civilized advancement."[197] The same spirit of technological optimism and confidence pervades the work of Joralemon, Rickard, Parsons, and other far less wealthy mining engineers who charted what they sincerely, if uncritically, saw as an epic history of human progress through modern copper mining. They credited the achievement to advances in scientific understanding of the natural world, to more powerful and efficient machines, and to improved means of measuring and controlling the complex business of reducing a mountain of rock to piles of fine dust. Above all, they gave credit to the mining and metallurgical engineers like Jackling, which is to say, to themselves. Justifiably proud of their accomplishments, to these engineers there seemed to be no natural or human obstacle so great that their intellectual and physical tools could not overcome it.

Yet, even as Jackling was celebrating the engineers as the masters of human destiny, the very success of their own ideas was fast breeding new problems. The subterrestrial mine had long sustained the engineers' faith in their abilities to understand and control environments. Now, as the subterrestrial increasingly gave way to the terrestrial mine, the environmental engineering challenges became ever more complex. Back in Butte, where a younger sibling of the Bingham Pit soon took root and began to grow, the engineers were shaping not only the destiny of humans but also that of thousands of Douglas fir trees, horses, hay fields, and countless other plants and animals in the Deer Lodge Valley. By midcentury, it no longer seemed so clear that the fate of any of them was in good hands.[198]

FIVE

The Dead Zones

Butte will never be a ghost city.

—Cornelius F. Kelley

Perhaps in the world's destruction it would be possible
at last to see how it was made.

—Cormac McCarthy

Walk through the older parts of Butte's Mount Moriah Cemetery, pause now and then to read the names and dates on the gravestones, and it will not be long before a pattern begins to emerge. On a late winter day in March, a brittle crust of snow still hides the yellowed cemetery grass. Deep drifts linger in the shade of the scattered pines where only the tops of the headstones are visible, poking out of their wells of snow like hard gray spring flowers. Bend and brush the snow away, though, and many of the grainy stone epitaphs will reveal a hard if not altogether surprising truth: men died young in Butte.

In the first few decades of the twentieth century, underground mining was the most dangerous industrial occupation in the nation, and Butte was home to one of the largest industrial mining operations in the world. In just the handful of years between 1916 and 1920, 410 Butte miners died underground. Thousands more were seriously injured. Frail human flesh was no match for fast and powerful metal machines, and safety regulations lagged behind the continual introduction of new dangers.[1] Butte has four main cemeteries. Roughly segregated by homelands and faiths, they are distant echoes of the proud and boisterous ethnic neighborhoods that once ordered the city. For some miners, their last trip through Butte was down the hill to Mount Moriah Cemetery on the flat valley floor. Most miners did not ask, or their families and friends did not wish, that their grave markers record the manner and place of their death. But the Butte writer and historian Zena Beth McGlashan knows of at least one exception: a headstone that marks the 1904 death of an Austrian immigrant to Butte named Michael Terzovich. The inscription on Terzovich's tombstone is much like the others in the cemetery, with the exception of the final words: "Killed in the Original Mine April 5 1904." A local newspaper account adds flesh to the stark words: on that Tuesday more than a century ago at 1,690 feet below the surface in the Original mine, an immense slab of rock fell and crushed Terzovich against the mine support timbers. He died a few hours later in Butte's St. James Hospital.[2]

Even with the March snow, it is easy to find the tall obelisk of Michael Terzovich's gravestone with its simple unadorned cross at the peak. Stand by the grave and look north and the hillside where Terzovich worked and died looms close above, almost as if the hill honeycombed with old mines were itself a giant tombstone for all its victims buried there. Today, decades after underground mining ceased, a handful of the old surface headframes remain, gaunt black steel reminders of the city's mining past. The city also remembers the minerals that came from beneath in the names of the cross streets that ascend the hill like the benches of an open-pit mine: Platinum Street, Gold Street, Silver Street, and Mercury Street. Copper Street holds a place of honor closest to the summit. Look a bit farther to the east and the big back wall of the Berkeley Pit offers an even more obvious reminder of Butte's mining past, one in which a sizeable section of the subterrestrial world in which Michael Terzovich once labored was gradually laid bare to the sky. When Terzovich died, the opening of the Berkeley Pit was still half a century in the future. But had he been able to work there in 1904, his odds of dying from a rock fall or most other common mining accidents would have been far less than in the Original mine. The pit could have saved Terzovich's life—though it might just as well have cost him his job, since open pits needed far fewer workers than underground mines.

If the Berkeley Pit later kept some miners from making a premature trip down the hill to the Mount Moriah Cemetery, it exacted many other costs instead. As the subterrestrial was made into the terrestrial, cemeteries were joined by other types of dead zones, both human and natural. In the first decades of the twentieth century, the Anaconda's engineers and managers had believed they could solve many of the pollution problems in the Deer Lodge Valley. The same modern tools of science and engineering that had allowed them to build the vast subterrestrial mines and sprawling Washoe smelter would provide the solutions. Company outsiders like Frederick Cottrell, Joseph Holmes, and John Hays Hammond agreed, and they and others associated with the Anaconda Smoke Commission worked to find or create the necessary technological fixes. Though they were sometimes arrogant in their overconfidence, they achieved some successes, and they might have achieved more had the mines and smelters of Butte remained static. Even as they were charting their best-laid plans, though, 350 miles to the south Daniel Jackling had been revolutionizing the very nature of the modern mining industry.

There had long been technological exchanges back and forth between Jackling's mass destruction mine and the mines and smelters of Butte and Anaconda. Recall that Jackling first witnessed the extraordinary selective power of flotation during a consulting trip to Butte in 1912.[3] Within a decade, he had made flotation the heart of his copper concentrating system. Jackling later repaid his debt to Butte by demonstrating the power of high-speed open-pit mining. When Anaconda began excavating the Berkeley Pit in 1955, the company followed a by then well-worn trail that Jackling had blazed a half century before. Paired with voracious new demands for copper in the postwar consumer society, this final convergence between the two great western mining regions created unprecedented stresses on the environmental and technological systems of Butte and the Deer Lodge Valley. Cornelius Kelley may have been literally correct in his optimistic prediction "Butte will never be a ghost city." But while Butte certainly survives to this day, its mines and smelter have left behind ghosts and dead zones that will haunt the city and the nation for centuries to come.

A FEARFUL SYMMETRY

There were early signs of the environmental disasters to come, though they were largely ignored at the time. In 1911, the Anaconda had agreed to "use its best efforts" to reduce or eliminate noxious smoke. The company accepted the oversight of a smoke commission dedicated to the task, and it made considerable progress in researching the pollution problems and developing improved controls. Nonetheless, between 1916 and 1918, the amount of dust, fumes, and gases released from the Washoe stack actually increased. The reason was simple. Accelerated demand from World War I encouraged the Anaconda nearly to double its rate of copper production. As the company now well knew, thanks to the years of environmental and technological study and measurement, when the Washoe smelter operated at full capacity the increased volume of smoke sped through the flue systems so quickly that much less noxious dust settled out.[4] Further, even if pollution controls like the Cottrell precipitators reduced the percentage of pollutants released into the atmosphere for any given ton of ore, the total tonnage of ore processed grew so quickly that the amount of pollution released could remain the same or even increase.

The earlier inadequate flue and stack system was supposed to have been remedied by the big new 585-foot stack completed in 1918. By that time, however, the Anaconda's adoption of flotation technology was working against the new stack's effectiveness. Just as Jackling was doing at Bingham, the Washoe engineers had begun replacing large parts of their concentrating system with flotation tanks. With flotation, the Anaconda metallurgists were able to recover an extraordinary (for the time) 91 percent of the copper from their Butte ores, versus an earlier recovery rate of only 78 percent. Unfortunately, successful flotation also required very fine grinding of the ore. As a result, it created far greater volumes of tiny particles of arsenic, cadmium, lead, and other pollutants. When these superfine dust particles were further processed by the smelter roasters and reverberatory furnaces, they were often too light to settle out in the big new dust chambers.[5] Even when the powerful Cottrell precipitators went online in 1923, the increased production of copper and fine dust undermined their effectiveness. The smoke commissioners were pleased to report that the Cottrell machines had reduced the output of arsenic from the Washoe smelter to a third of its previous level. That remaining third, however, still amounted to as much as twenty-five tons of arsenic per day. The pollutants that evaded both the dust chambers and the Cottrell machines would only increase with greater throughput of copper ore.[6]

At base, the seemingly insatiable American appetite for copper was working against the development of a viable solution to the Anaconda smoke problem. When the Washoe smelter first opened in 1901, Americans consumed about 214,000 tons of copper per year. Though there were some years when demand slackened slightly, overall American consumption continued to increase until it reached a 1918 high of 998,000 tons—more than four times what it had been in 1901. With the end of World War I, consumption dropped sharply, and it continued to fluctuate during the stormy economic years of the early 1920s. But by 1923, the economy was again on the upswing and Americans were using more copper than ever. On the eve of the Great Depression that would eventually force many copper mining operations to shut down, consumption reached a record high of 1,334,000 tons per year—six times the amount consumed when the Washoe opened in 1901. The often volatile price of copper did not always climb as steadily as demand, but after a World War I high of about twenty-nine cents a pound, prices averaged a low but sustainable fifteen cents during the economic boom times of the 1920s. Indeed, copper prices stayed so low in part because Jackling and others were developing huge new deposits of ore. Production expanded to meet (or even exceed) demand, kept prices low, and thus encouraged even more copper consumption.[7]

At Anaconda, managers and engineers increased production both to profit from high wartime prices and to meet the growing consumer demands of the 1920s. With these increased ore volumes and the adoption of flotation, the challenges of engineering a solution to the pollution problems in the Deer Lodge Valley became considerably more daunting and costly. Thanks to the company studies over the previous decade, the Anaconda engineers and managers understood fairly well what this increased production meant in terms of damages to forest, crops, and livestock. Research on the complex interactions of the smoke and precipitation patterns also suggested that during future periods of drought the damage would become even worse, raising the specter of new suits from farmers and ranchers.[8]

Faced with these realities, there was still much the Anaconda engineers could have done to alleviate the smoke pollution problems. The company could have spent more on researching and adopting pollution control systems, production rates could have been cut back during adverse weather conditions, and even the use of flotation could have been limited—there were many possibilities, though all would entail some loss of profits. Other smelters around the nation facing similar problems did adopt some of these measures, though usually only as a result of legal pressures. A few mines and smelters simply shut down.[9] At this point, however, the Anaconda increasingly retreated from its earlier attempts to solve the smoke problem in favor of a solution that essentially allowed the company to ignore the Deer Lodge Valley environment altogether.

Beginning in the early 1920s, the Anaconda embarked on a series of land swaps with the federal government. The company gave the Department of the Interior undamaged forest lands it owned around the state. In exchange, the government gave the Anaconda title to large tracts of national forest land near the Washoe. Between 1921 and 1935, these land swaps made the Anaconda the owner of almost all of the national forest lands that had been (or might be) damaged by the smelter smoke. During this same period, the company had quietly continued buying up farmland in the valley, a tactic it had been using since the smoke litigation first began in 1902. After years of struggling with the Anaconda and the smoke problem with little to show for it, many of the farmers were now willing to sell out. Where the Anaconda did not buy simple title to the land, it was often able to purchase "smoke rights" in which the owner agreed not to sue for any damages the smelter smoke might cause to the land.[10] By the early 1930s, the Anaconda either owned or had the legal right to pollute almost all the farm and forest land around the Washoe likely to be harmed by the smelter smoke. As one historian of the land deals puts it, the Anaconda had gained "sovereign authority for the right not only to pollute the air in the Anaconda region but for the right to pollute or damage the public domain whenever, wherever, and however it chose."[11]

This, then, was the final outcome of the earlier confident expectations that modern science and engineering could and would solve the smelter smoke pollution problem: not a balanced integration of the technological system of the smelter with the ecological system of the valley, but rather an almost complete denial of the Deer Lodge Valley's ecology—a retreat from nature. The Anaconda had created a sovereignty of smoke, an empire of pollution. Attempts to understand the complex interactions between the technological and ecological no longer seemed necessary. Once the smoke passed beyond the high windy rim of the Washoe stack, the company believed, it could safely ignore where it went and what it did.[12] As one historian of the pollution problem rightly notes, at this point the "environmental consequences were apparently accepted as a given, much like the ore deposit in the Butte hill from which the mineral values were extracted."[13]

Given the Anaconda's initial good faith efforts to solve the smoke problems and integrate its mining and smelting technology with the local ecology, how can we account for this belated retreat from nature? A number of historical factors offer some explanation. The increasingly probusiness climate of the 1920s tended to eclipse the previous Progressive Era ideals of conservation and corporate responsibility, which, while far from ecological in scope, nonetheless encouraged a less irresponsible treatment of natural resources. Historians have also pointed out that professional engineers during this period lost much of their previous independence and sacrificed higher ideals of societal service for the promise of secure corporate employment.[14] Perhaps most important, though, the decade of the 1920s was a period of economic optimism, a time when the astounding powers of mass production were promising a new age of inexpensive consumer abundance in which all Americans could afford electric lights, refrigerators, and automobiles. Mass consumption had arrived, along with its promise of a democracy of goods that would end class disparities and conflicts and create a more peaceful and equitable America. In such an exuberant economic and social climate, few paused to consider the question of where all the raw materials would come from to build this new consumer utopia.

Had Americans bothered to entertain such a question, they would have found the answer in places like Butte where the ruthless logic of modern economic efficiency had yet to reach its full illogical extent. When the Anaconda switched to flotation technology at the Washoe, greatly increasing the amount of dust pollution, it did so in part because it would allow the company to profitably mine lower-grade copper ores in Butte it had previously avoided. Indeed, the flotation system was so efficient at separating copper and other metals from masses of worthless waste rock that it eventually became cheaper for the Anaconda to adopt the technology of mass destruction that Daniel Jackling had pioneered at Bingham. In 1955, the company commenced open-pit mining, using trucks instead of trains to transport the ore but otherwise replicating most of the elements of Jackling's Bingham system. The Berkeley Pit freed the engineers from the constricting shafts, complex ventilation systems, and other underground technologies their predecessors had pioneered a half century earlier. Giant power shovels replaced miners, each machine capable of moving tons of blasted ore in minutes. Just as Jackling had shown at Bingham, it had become easier and more efficient to dig down from the surface, blasting out all of the variable Butte ore regardless of grade and sending it over to the Washoe, where the flotation tanks would pick out the small amounts of metal. With the extremely low-grade ore coming out of the Berkeley Pit in the 1950s, the flotation tanks kept only about a ton of material for every nineteen tons of ore processed. Of that ton of concentrate, the Washoe smelter drove out further impurities like sulfur, arsenic, and other heavy metals. The big flues and Cottrell precipitators captured some, but much still escaped into the atmosphere. The remaining nineteen tons of waste, which still contained significant levels of heavy metals, went over to the rapidly growing Opportunity tailings ponds to the east of the smelter.[15]

Berkeley Open Pit Draws Thousands of Tourists

BUTTE
THE WORLD'S GREATEST MINING DISTRICT

17,500 Tons of Copper Ores Per Day To Come From New Project on Butte Hill

This vast expanse before you, the Berkeley Pit, is one of eight open pits operated by The Anaconda Company throughout the world, namely . . . . The Chuquicamata in Chile, Yerington and Leviathan in... In the Reduction Plant the ore goes to a central crushing plant from which it is conveyed to the Concentrator where it is further crushed and ground by means of rolls and ball mills until it reaches... This finished copper is then shipped to American Brass Co. plants in Connecticut and to Anaconda Wire & Cable plants throughout the country.

15. A free publicity flyer celebrates Butte's recently opened Berkeley Pit as a tourist destination. The new pit operated on the principles of mass destruction Jackling had pioneered at Bingham fifty years before, although the Anaconda used big dump trucks instead of trains to move the ore out of the pit. *Anaconda Copper Trailsman, August 15, 1956.*

With the immense amounts of waste produced by the combination of floatation technology and mass destruction mining of low-grade ore, the challenges of successfully engineering a balanced environmental and technological system in the Deer Lodge Valley continued to grow. From a simple environmental perspective, the Anaconda should have simply shut down its operations when the higher-grade ore first began to give out in the 1930s. However, this would have entailed the loss of a significant source of copper for the approaching world war and the exploding postwar American consumer society. Instead, the company, like so many others involved in twentieth-century natural resource extraction, embraced a narrative of abundance that essentially denied the existence of almost any limits to mineral production, whether environmental or physical.[16] Shortly after opening the Berkeley Pit, the Anaconda proclaimed that "the full geological limits of its resources are still to be determined" and the "ore reserves in Butte today are greater than the total tonnage extracted in all the years since 1864."[17] True enough, but only if the company was permitted to pollute with near impunity. If the Anaconda wished to demonstrate that there were no physical limits to the amount of copper ore in Butte, it demanded, and was given, the power to reject almost any limits to its destruction of land and production of waste in the Deer Lodge Valley.

The result was the creation of two massive "dead zones," areas of such intense environmental exploitation that the technological overwhelmed the ecological and rendered the landscapes nearly sterile. The first was the Washoe smelter site, an immense area of dead forestlands, ruined farms and ranches, towering black slag heaps, and lifeless tailings ponds that stretched for miles. The second was the Berkeley Pit itself, a gaping yellow and gray amphitheater of rock where few living things remained other

16. Despite the new flue system and banks of Cottrell precipitators, increased production and the adoption of flotation techniques meant that large volumes of dangerous pollutants still escaped from the Washoe stack. The result was an immense dead zone of toxic tailings and barren hills and valleys where scarcely any vegetation could grow. *Courtesy Montana Historical Society Photograph Archives, Helena.*

than the human workers with their big machines and explosives. Together, these two dead zones helped provide the copper that created and sustained the celebrated American way of life.

The essential environmental and technological roots of mass production and consumption have long been obscured in modern industrial societies. A stylish 1953 Ford Victoria sedan, filtered through multiple layers of advertising and cultural meaning, might appear as a symbol of individual personal expression and status, or perhaps of the astounding productivity of Detroit's factory system. Few think to trace the Ford's copper radiator and yards of electrical wire back to the envirotechnical systems and dead zones in Butte and Anaconda that made them possible. As if in a dark subterrestrial mine, the true environmental roots of the car remained hidden, the consequences of its mass production and consumption obscured in shadows. The Ford Victoria, after all, seemed to be an artificial, human-made artifact, one far removed from the organic natural world most people think of when contemplating "the environment."

Part of the usefulness of thinking in envirotechnical terms lies in the collapsing of precisely these distinctions. The production of cheap and abundant Ford automobiles would have been impossible absent the mining and smelting systems created by engineers and managers at Butte and Anaconda, as well as at hundreds of other open-pit mining operations around the world. In this light, the Berkeley Pit can be seen as the purest expression of the high modernist ethos of maximum productivity, control, and efficiency (the very same principles that made cheap Ford cars possible) taken to their "logical" ends. The engineers' domination of the subterrestrial world was extended into the terrestrial world as the underground mine literally became the surface, the seemingly inorganic fused inextricably with the organic, and the technological merged into the natural. Faced with the intractable complexity of the terrestrial environment, the engineering and scientific methods of the Anaconda engineers proved inadequate. As a result, the Anaconda engineers and managers largely abandoned their efforts to create a more balanced environmental and technological system of mining and smelting, instead choosing to extend the radical simplification of the mine to a large section of the Deer Lodge Valley. The result was yet another "dead zone" that in many ways mirrored that of the pit, creating a rough symmetry of mass destruction embedded into the landscape itself.

The Berkeley Pit and the Washoe smokestack thus offer mute but enduring evidence of the essential envirotechnical nature of high modernist mining, and evidence too of the dangerously misleading illusion that the subterrestrial environment can remain separate from the terrestrial, the technological from the natural, and, indeed, the Ford Victoria from the nation's largest Superfund site. Should tourists speeding through the Deer Lodge Valley today in their modern Fords, Toyotas, and Volkswagens decide to turn off the interstate and drive the few miles west to take a closer look at the looming tower of the Washoe smelter stack, they will discover that it is now part of the "Anaconda Smoke Stack State Park." The site's brief interpretive material, available at a viewing site in an Anaconda city park, will provide them with little sense of the complex historical forces that created the stack and the damaged lands surrounding it. Still, some inkling of the story may yet come through when they realize that the Anaconda Smoke Stack site has a notable difference from most of Montana's bucolic state parks. Appropriately enough, it is a park visitors are forbidden to enter, an environmental dead zone marked by a 585-foot high tombstone.

Such were the ambiguous memorials to Jackling's system of mass destruction, unintended and unwelcome reminders that the American way of life had come only at steep cost. Indeed, the basic principles of mass destruction had surprisingly deep roots in the emerging culture of modern America.

THE CULTURE OF MASS DESTRUCTION

On the night of February 25, 1945, a sortie of American B-29 bombers attacked Tokyo, dropping tens of thousands of incendiary bombs over a one-square-mile area at the heart of the city. When ignited, the bombs' mixture of jellied gasoline and magnesium burned slow but hot and stuck tightly to whatever it hit. The result, in the words of one of the B-29 pilots who later wrote about the bombing, was a literal "hell on earth." As Tokyo's mostly wooden buildings exploded in flames, crowds of men, women, and children fled the fires, though few would find safety as giant firestorms roared through the city. Many sought refuge in the rivers and canals that snaked through Tokyo, only to be subsequently crushed or drowned as wave after wave of terrified human beings jumped in after them. Others were asphyxiated as the inferno consumed all the oxygen in the air. Eventually, even the river water itself began to boil.[18]

The first half of the twentieth century was the bloodiest in history. Aerial bombing, incendiary weapons, atomic bombs, and other modern weapons of mass destruction killed thousands in the course of only a few hours or seconds, leaving nothing behind but miles of rubble and skeletal steel frames. As was noted at the start of this book, the phrase "mass destruction" seems an apt description for Jackling's technology of open-pit mining, even though many today equate these words with weapons of war and terrorism. Obviously, there are important differences between the creation of the Bingham and Berkeley pits and the mass destruction of cities during World War II. At a deeper level, though, these different types of dead zones share some striking similarities, suggesting they both rose from the same suspect technological and moral logic that lay at the heart of a modern culture of mass destruction.

In the last nine months of World War II, American bombers dropped about 159,000 tons of bombs on Japanese cities. In one attack on Tokyo on the night of March 9–10, 1945, more than three hundred B-29 bombers flattened sixteen square miles of the city and killed an estimated eighty-three thousand people. It was, one reporter believed, "the first all-out effort to burn down a great city and destroy its people."[19] This was not entirely accurate. The previous month, two waves of British bombers dropped several thousand tons of incendiary and explosive bombs on the German city of Dresden. The resulting firestorms leveled much of the beautiful old medieval city and killed perhaps as many as forty thousand civilians. Indeed, even as early as February 1941, the British Air Staff had decided to bomb built-up residential areas as a means of undermining "the morale of the enemy civilian population and in particular of industrial workers."[20]

To their credit, American military leaders initially rejected the use of such indiscriminate carpet bombing of civilian populations in favor of so-called precision targeting of military or industrial targets. Even with the much-vaunted Norden bombsight, however, precise aerial bombing was difficult given the exigencies of combat, unpredictable weather conditions, and poor visibility. By the latter years of the war, the Americans too began indiscriminate mass bombing of enemy cities and civilian populations, as the near annihilation of Tokyo and many other Japanese cities amply demonstrates.[21] In this light, the decision to use atomic weapons on Hiroshima and Nagasaki in August 1945 was hardly surprising. At the time, a handful of individuals did question the morality of incinerating thousands of Japanese women and children in order to end the war quickly and spare the lives of American soldiers. But by the summer of 1945, most of the nation's civilian and military leaders had already become accustomed to the indiscriminate killing of civilian populations. Curtis LeMay, the head of strategic air operations during the final months of the war, later defended the use of the atomic bombs by pointing out that "we scorched and boiled and baked to death more people in Tokyo on that night of March 9–10 than went up in vapor at Hiroshima and Nagasaki combined."[22]

LeMay underestimated the death toll of the two atomic bombs, but his basic point was valid. While today we may think of the atomic bomb as the prototypical weapon of mass destruction, well before August 1945 there were plenty of other contenders for the title. The phrase itself seems to have first appeared in print on December 28, 1937, in a page nine story in the London *Times*. The story was mostly a transcript of the archbishop of Canterbury's Christmas radio broadcast on the topic of "Christian Responsibility." The year 1937 had seen much carnage from aerial bombardment. In April, Hitler's Luftwaffe flattened the heart of the Spanish town of Guernica, killing more than 1,650 civilians and inspiring Pablo Picasso's famous mural. The following August brought news of an equally shocking Japanese bombardment of the largely civilian population of Shanghai. Deeply disturbed by such indiscriminate and "appalling slaughter," the archbishop struggled to capture its significance for the future of humanity, and thus an ugly new phrase entered into the English language. "Who," the archbishop wondered, "can think without horror of what another widespread war would mean, waged as it would be with all the new weapons of mass destruction?"[23]

In his brilliant examination of the American development of modern insecticides, *War and Nature*, the historian Edmund P. Russell argues that war and the human domination of nature coevolved during the twentieth century. "The control of nature expanded the scale of war," Russell notes, "and war expanded the scale on which people controlled nature."[24] Responding to European threats during World War I, the American Chemical Warfare Service developed new poisonous gases and chemical weapons—another early class of weapon of mass destruction whose use sometimes affected civilians and soldiers alike. Following the war, the Chemical Warfare Service tried to adapt its poisons to a new peacetime "war" on insect pests, a crucial step toward linking warfare on humans to warfare on nature. These technological, ideological, and organizational exchanges grew even more pronounced during World War II as the doctrine of "total war"—which made little distinction between civilian and military targets—gained wide acceptance on all sides of the conflict. Some American strategists believed that a fanatical Japanese army would never surrender and must be annihilated, much as an exterminator might annihilate a nest of ants. Such analogies between humans and nature, Russell argues, helped justify the use of weapons of mass destruction like incendiary and atomic bombs. "In the era of mass production and mass attack," Russell notes, "thinking of human enemies as masses of insects was useful—especially if one believed war should aim for extermination."[25] In the years following World War II, Americans again adapted the weapons and ideologies of war on humans to the control of nature, using DDT and other new chemicals in an ultimately futile and dangerous attempt to eradicate insect pests like mosquitoes and flies from entire cities, or even entire states—a "total war" of annihilation.[26]

As Russell's provocative work suggests, it is no coincidence that twentieth-century industrial societies simultaneously gave birth to the mass destruction of both humans and nature. In the case of pesticides, the technologies and ideologies of chemical warfare on humans and insects were mutually reinforcing and in constant interchange. Similar parallels existed between the mass destruction of nature through open-pit mining and the mass destruction of humans through aerial bombing, poison gases, and other new weapons of war. Both methods of mass destruction permitted the use of powerful but often crudely indiscriminate new technologies, from dynamite and other modern chemical explosives to the caterpillar-equipped tanks and steam shovels capable of maneuvering over rugged uneven terrain. Both emphasized the speed and efficiency with which they could achieve their goal, whether that was the reduction of a mountain or a city to a pile of rubble. As Vannevar Bush, a leading science and engineering adviser during World War II, argued in 1944, the use of incendiary weapons against the Japanese offered "a golden opportunity of strategic bombardment in this war—and possibly one of the outstanding opportunities in all history to do the greatest damage . . . for a minimum effort."[27] Or, as another physicist regretfully described his wartime scientific efforts, "I sat in my office until the end, carefully calculating how to murder most economically another hundred thousand people."[28]

Above all else, both the mass destruction of humans and of mountains achieved their increased speed and efficiency through their shared principle of nonselectivity. Just as Jackling's open-pit mine abandoned the slow but precise methods of selective underground mining of ore, so too did the strategy of mass aerial bombardment abandon the slow but somewhat more humane method of selective killing—a method that at least attempted to distinguish between military and civilian targets, combatants and noncombatants. Of course, for all the damage it wreaked on the natural environment, Jackling's system of mass destruction mining did not raise moral issues quite so deeply troubling as those surrounding the mass destruction of entire cities of human beings. Nonetheless, perhaps the most disturbing similarity between the two methods of mass destruction was the way in which their advocates used the claim of technological inevitability to avoid discussions of the moral or environmental issues that lay outside the realm of mere efficiency and feasibility. Jackling and others argued mass destruction mining was the only logical, efficient, and thus seemingly inevitable means available to obtain cheap and abundant copper. Jackling likely would have understood and applauded Vannevar Bush as he analyzed the numbers and concluded that the cheapest and quickest means of winning the Pacific war lay with mass incendiary bombing of civilian populations.

In their mutual embrace of such narrowly proscribed ideas of rational efficiency, the advocates of both types of mass destruction also often demonstrated their belief in a certain strain of technological determinism: the idea that any new technology would follow its own internal logical and inevitable course of development. In this view, technologies were autonomous and essentially immune to the petty efforts of humans to control and direct them. Such beliefs could only have gained credence in an era in which many had come to believe that technological systems had overwhelmed all other forces that had once been thought to drive the course of human history, such as the natural environment, ideology, politics, or religion. Yet, as many historians have shown, just such a belief became pervasive during the twentieth century, with its uncritical faith in technological efficiency, rationality, and control.[29] No doubt many may have truly believed that technologies were autonomous and beyond their control yet such ideas also offered convenient justifications for the environmental and human dead zones that resulted. These were the principles at the heart of the modern culture of mass destruction: that the annihilation of a mountain or a city must be the inevitable result of any efficient calculation of costs and benefits.

In both cases, of course, the truth was far more complicated. To be sure, Jackling and the Anaconda were correct when they claimed their efficient open pits would provide cheap copper to fuel a burgeoning consumer economy, just as the advocates of aerial bombing were correct that it would provide a cheap way of killing the enemy and winning wars. Nonetheless, there were always alternatives, always moral decisions to be made. They may not have been easy or convenient choices, but they did exist.[30] As one historian of the American bombing campaign in World War II notes, most of the strategists were too "immersed in technical deliberations and problem solving" to seriously ponder the morality of their actions.[31] Likewise, Jackling and those who came after him were too obsessed with the drive to profit from feeding the American demand for cheap copper to think deeply about the negative consequences of mass destruction mining on farmers, ranchers, forests, and rivers. In both cases, increased speed and efficiency were achieved only by putting to use powerful but blunt new tools of the day, whether those be heavy steam engines or heavy bombers. In so doing, the true costs for cheap copper and cheap military victories were shifted to the mass destruction of mountains and of cities, creating the illusion of efficiency by exacting a disturbingly high price elsewhere.

As the archbishop of Canterbury said during the same sermon in which he warned of the impending horrors of war by weapons of mass destruction, such insanities could only exist among a people who acquiesced to being "absorbed in the vast machine of modern civilization."[32] The archbishop's comments were more accurate and prescient than he could have known. Industrial humans had built the modern world in such a way as to make its own destruction at times seem logical, efficient, inevitable, and even desirable. At their roots, the mass destructions of nature and of humans were nourished by many of the same technologies, ideas, and myths. Today, of course, more than a few historians and other commentators have suggested that the mass slaughter of hundreds of thousands of civilians during World War II was morally inexcusable. For obvious reasons, the mass destruction of nature has not inspired equivalent condemnations, although its consequences continue to become more severe and apparent. However, it is one thing to question decisions made during a bitter war waged more than half a century ago, and quite another to face squarely the problems of a modern system of mass destruction that continues to be essential to providing the world with cheap raw materials and energy from the earth.

The dilemmas presented by mass destruction mining went to the very heart of the modern American social, economic, and political order: the culture of mass consumption. It was a connection the mining companies became increasingly sophisticated at pointing out. The culture of mass destruction, their campaigns suggested, formed the material bedrock of the culture of mass consumption.

Mass Destruction

In the summer of 1957, the Anaconda Company ran an intriguing advertisement in the *Saturday Review*, a popular magazine of American culture. The ad encouraged readers who might have already planned to visit more traditional American tourist destinations that summer to also consider spending some time on a tour of "America the BOUNTIFUL." In particular, the productive environs of Butte, Montana, were "within easy driving distance" from Yellowstone National Park, so why not take a little detour and see the "America that's *Bountiful* as well as Beautiful." True, Butte had no charming black bears begging for potato chips and hot dogs, no totemic tourist sites like Old Faithful geyser or the Grand Canyon of the Yellowstone. But what the city did have, the ad explained, was more than 2,700 miles of mine tunnels and shafts that tapped the "seemingly inexhaustible mineral wealth of 32 square-mile area whose output increases year after year." Above all, as the half-page photo accompanying the ad emphasized, at Butte "you can stand on a ledge 500 feet above the floor of the new Berkeley 'canyon' "—the ad's picturesque term for the open-pit mining operation the Anaconda had begun a few years earlier.

If the Anaconda's Berkeley "canyon" may have lacked some of the sublime beauty of Yellowstone's Grand Canyon, it was undoubtedly far more "bountiful"—at least if bounty was measured by vast tons of lean copper ore transmuted into endless spools of electrical wire and miles of gleaming pipe.[33] Transmutation, despite its connotations of magic or miracle, seems the right word here, as there was something of the miraculous happening in Butte. As the 1957 magazine ad rightly suggested, Butte was the perfect place to "see how the Anaconda Company mines the metals so essential to the nation's economic strength."[34] While at first blush these words sound very sober and down-to-earth, the hint of the miraculous lay in the ad's subsequent claim that "seemingly inexhaustible mineral wealth" could be found in a decidedly finite thirty-two-square-mile region around the city. Combine these two phrases—"the nation's economic strength" and "inexhaustible mineral wealth"—and the deeper significance of Butte begins to emerge. Butte was symbolic of a bountiful America not just because it promised inexhaustible extraction of an essential industrial mineral but also because this endless supply of copper made it possible for Americans to indulge in inexhaustible consumption. As postwar Americans emerged weary and battered from nearly two decades of economic deprivation, they had rushed to embrace the liberating joys of

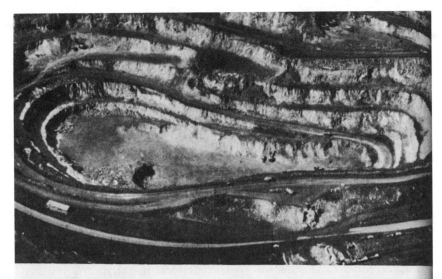

THIS SUMMER, PLAN ALSO TO

see America the BOUNTIFUL

VISIT BUTTE, MONTANA —"THE RICHEST HILL ON EARTH"— AND SEE HOW THE ANACONDA COMPANY MINES THE METALS SO ESSENTIAL TO THE NATION'S ECONOMIC STRENGTH.

The
ANACONDA
Company

Seeing the America that's *Bountiful* as well as *Beautiful* can be an interesting part of your vacation itinerary. This summer if you plan to take in the scenic spectacle of Yellowstone Park, you will be within easy driving distance of Butte—home of The Anaconda Company and for over 70 years one of the world's most spectacular producers of copper and other important metals.

The American Brass Company
Anaconda Wire & Cable Company
Andes Copper Mining Company
Chile Copper Company
Greene Cananea Copper Company
Anaconda Aluminum Company
Anaconda Sales Company
International Smelting and Refining Company

Butte sits atop a veritable honeycomb of tunnels, shafts and passageways totaling some 2700 miles in length. More than 6500 of its residents are employed by Anaconda in extracting the seemingly inexhaustible mineral wealth of a 32 square-mile area whose output increases year after year.

On guided tours you can go underground in the Kelley Mine and see how 15,000 tons of ore a day are mined by the ingenious block-caving method. You can stand on a ledge 500 feet above the floor of the new Berkeley "canyon," illustrated above. This is a giant open pit operation now nearing an output of 17,500 tons of ore daily. Or, if you wish, you can "do" Butte, the city and its mines, on your own schedule. The Anaconda Company will be glad to send you a do-it-yourself map and guide to a memorable tour of the famed "Copper Trail." Simply write the Company at Butte, Montana.

The Anaconda Company extends a cordial invitation to inspect "The Richest Hill on Earth" and believes your stay will give you a new pride and appreciation of your America the Bountiful.

17. The Anaconda ran this advertisement in the summer of 1957 to encourage tourists to visit the company's new Berkeley Pit. By linking mass destruction mining to the emerging culture of mass consumption, the ad suggests that modern technology could guarantee a nearly inexhaustible source of natural raw materials for the emerging "America the Bountiful." Saturday Review, *July 27, 1957, inside front cover.*

consumerism, to become once more the chosen "people of plenty."[35] The Anaconda's confident promise of an enduring "America the Bountiful" was really the promise of endless rows of refrigerators with copper coils, sales lots full of cars with copper radiators, and acres of new homes with copper plumbing and wires.

Even though no one may have dared substitute "Oh bountiful for spacious mines" for the familiar words of the popular patriotic anthem, the Anaconda Company's ad nonetheless nicely captures the postwar American determination to sustain strong economic growth though mass production and consumption. As Gary Cross and other historians demonstrate, in the latter half of the nineteenth century mass consumerism became more central to the "American Way of Life" than did almost any of the nation's other political or cultural ideals. "Consumerism," Cross suggests, "was the 'ism' that won." In an era of international cold war and domestic red scares, American leaders saw mass consumerism as a means of fostering stability and unity at home while helping to undermine the appeal of the Communist alternative around the globe.[36] Only two years after Anaconda encouraged Americans to "see America the bountiful," Vice President Richard Nixon would spar with Soviet Premier Nikita Khrushchev in the famous "Kitchen Debate." "Would it not be better to compete in the relative merits of washing machines than in the strength of rockets?" Nixon suggested. Khrushchev agreed, and though the conversation quickly shifted back to whose generals had the biggest missiles, the point about America's astonishing ability to manufacture immense quantities of popular consumer goods hit its mark. Midcentury America, the Kitchen Debate suggested, had become "a consumer's republic."[37]

It is no wonder, then, that the Anaconda, Utah Copper, and other producers of primary raw materials would want to associate their industries with this miraculous new consumer cornucopia. After all, copper was an essential component of the electrically powered washing machines Nixon boasted of so proudly. The Anaconda's "See America the Bountiful" ads in the 1950s were just the latest gambit in an ongoing public relations and marketing campaign that copper mining companies had been pursuing for decades. In part, such campaigns served simply to improve the public image of copper mining, curry political favor, and encourage expanded use of copper products. But copper mining's postwar public relations campaigns also began to redefine the very meaning of the technological landscape of mining. Through ads and tourism, the major copper companies attempted to link what had previously been rather straightforward places of industrial extraction with the exciting dynamism represented by the burgeoning new culture of consumption of the 1950s. This is not to say that these links were not genuine; to the contrary, in a time when modern technological civilization seemed increasingly disconnected from its natural foundations, the ads provided a valuable reminder to Americans about the first source of their postwar plenty. Not surprisingly, however, the companies neglected to point out that the productive miracles achieved by their giant open-pit copper mines often came at a high human and environmental cost. Thus, even while illuminating the essential links between consumerism and nature, the public relations campaigns simultaneously obscured deeper and less attractive realities of modern mass destruction mining.

Well before World War II, the managers of the Anaconda, Jackling's Bingham mine, and other big American copper companies realized that they had much to gain, both economically and symbolically, by illuminating the paths from mine to mill to consumer. During the period of depressed demand for copper following World War I, the Copper and Brass Research Association, an industry trade group, successfully spurred copper consumption through advertisements and pamphlets encouraging architects and building contractors to use more copper. By 1929, the association boasted that it had increased the use of copper in building fortyfold, and copper appeared more frequently in building facades, doorways, and portals.[38] In the early 1920s, the Anaconda Company even adopted a telling new corporate slogan: "From Mine to Consumer." The primary impetus for this new slogan was the increasing vertical integration of the copper mining company with copper, brass, wire, and cable manufacturing subsidiaries, yet it was also indicative of the public relations and marketing campaigns to come that would deliberately link extraction and consumption.

During the depression years of the 1930s, the American copper mining industry again suffered sharp declines, and little effort was made to promote what would have by then been a foolish assertion that copper production was the foundation of a vanished American prosperity. World War II, however, jolted the feeble industry back to life as demand for copper-containing bullets, airplanes, tanks, and all manner of electrical products soared. The war also made apparent the nation's growing dependency on foreign mineral supplies, as did the subsequent onset of the cold war and its battle by proxies in Korea. Fearful that the nation could run short of critical minerals, the federal government attempted to spur domestic mining through such postwar programs as the Defense Minerals Exploration Administration. Partly as a result of these federal efforts, by the

1950s the domestic mineral industry was widely recognized as essential to America's continued economic vitality and national security. In 1952, an Anaconda pamphlet asked, "Will the U.S. run short of metals?" Citing the President's Materials Policy Commission, the pamphlet suggested the answer was yes—unless industry increased production at home and secured supplies imported from abroad. The Anaconda promised it had already begun to "expand metal production and to find more efficient ways to fabricate and use metals."[39]

It was in this already positive economic and public relations climate for mining that the major American copper companies began to link the meaning of their technological landscape to the emerging consumer-driven economy of the postwar nation. As Lizabeth Cohen, Alan Brinkley, and other historians have recently argued, during the New Deal, liberal economic theorists began to stress the importance of robust consumer demand in maintaining economic health.[40] By the 1950s, the maintenance and continuous expansion of this consumer economy was viewed by many analysts as one of the chief responsibilities of a strong federal government. Democrats and Republicans alike embraced this centrist liberal theory of a consumer economy, in part because it seemed to offer a painless means of maintaining economic growth without expanding social welfare programs. Even the Republican president, Dwight D. Eisenhower, praised the benefits of the "American system" in which a "working man can own his own comfortable home and a car and send his children to well-equipped elementary and high schools and to college as well."[41] Simultaneously, many progressive politicians abandoned their earlier attempts to use political power to achieve more even distribution of wealth, and instead hoped to reach the same ends by rapidly expanding the consumer economy. This "Consumer's Republic," as Lizabeth Cohen calls it, "promised the socially progressive end of economic equality without requiring politically progressive means of redistributing existing wealth. . . . [A] growing economy built around the dynamics of increased productivity and mass purchasing power would expand the overall pie without reducing the size of any of the portions."[42]

Mass consumerism, then, was believed to be the miracle cure for all manner of potential economic and social ills. So long as ever more Americans kept buying ever more cars and refrigerators, the wolves of depression and class conflict could be kept from the door. Yet even allowing for increases in efficiency, substitution, and recycling of materials, an ever expanding consumer economy obviously necessitated an ever expanding supply of raw materials. Mass consumption demanded mass extraction—of timber, oil, copper, and a range of other raw materials. Cohen and other historians of American mass consumerism have done little to illuminate the extractive roots of this extraordinary postwar political economy. Mid-century copper mining giants like the Anaconda and Kennecott, however, were understandably eager to make the connection. The copper mining and manufacturing companies did so in two distinct ways. First, they educated consumers about the many uses of copper in an array of essential and popular postwar industrial and consumer items. Second, they portrayed mining operations themselves, and particularly open-pit mines, as the eternal foundations of economic prosperity, selectively educating the public about Jackling's mass destruction process that underlay the better-known phenomena of mass production and mass consumption. This is the context in which we must understand the "See America the Bountiful" campaign and the transformation of raw-boned industrial sites like Bingham and the Berkeley Pit into some of the more unlikely of American tourist destinations.

A 1944 advertisement from one of the Anaconda's principal manufacturing subsidiaries, the Anaconda Copper & Brass Company, offers an early example of the first of these approaches: the campaign to more closely link copper mining and manufacturing with consumer goods. In this ad the claim that "Copper Serves . . . Every Minute . . . Everywhere" reiterates earlier industry efforts to emphasize the centrality of copper to the American industrial economy. Illustrations of industrial uses of copper dominate, with a line connecting a shiny red copper ingot to a train, freight ship, airplane, and electric power plant. Significantly, though, even at this early date the ad also includes a drawing of a cozy two-story cottage, clearly suggesting the importance of copper to the average American homeowner. Even more telling, the text argues that copper "is symbolic of the American way of life," which the ad suggests can be defined by the widespread availability of comfort, leisure, and learning. Copper also provides these benefits through consumer goods, as "there's copper in your radio . . . copper and copper alloys in your refrigerator, plumbing, and heating equipment"—all the modern technological means of providing comfort and leisure.[43]

As is the case here, many early ads mentioned consumer uses of copper but primarily stressed its value to industry. This reflected widespread fears within the industry that demand for copper would collapse in the postwar years. Since military uses for copper took priority during the war, manufacturers of nonessential consumer products had been forced to substitute other materials like aluminum. Copper executives worried that some might not switch back to copper after wartime shortages ended. To encourage producers to return to the fold, Anaconda Copper & Brass ran a series of ads with the theme "Nothing serves like copper."[44] These ads argued that from automobile radiators to stoplights to airplanes, copper was still the best material available in terms of reliability and durability. One 1944 ad asserted, "Copper and copper alloys have been proved essential over and over again both in old fields and many new ones," thus making it "the key metal in so many postwar plans."[45]

Even during the war, though, copper companies still recognized the need to educate consumers about the many uses of copper. As the "copper serves" ad suggests, the importance of using copper in household plumbing was an especially common theme. One 1945 Anaconda Copper & Brass ad features pictures of four stylish and spacious houses and encourages consumers to "learn how little it will cost to rustproof any of these post-war homes."[46] Copper's natural resistance to corrosion, the ad argues, made it the prudent choice for building a safe and enduring home. That the Anaconda ads emphasized stereotypical suburban family houses with large yards and spacious floor plans was no coincidence. Suburban homes would be the single most important consumer item of the postwar period. Annual housing starts exploded after 1945, growing from a modest 142,000 in 1944 to a staggering 2,000,000 in 1950.[47] The phenomenon of "mass suburbia" was built on mass extraction of raw materials and mass destruction mining—no other single consumer purchase placed a greater demand on supplies of timber, brick, concrete, and, of course, copper, than did a new house. In this light, these advertisements were effective attempts to link copper mining and production with the most important consumer item of the period, and by association all of the social, cultural, and political benefits homeownership was believed to confer.

New homes were not all that Americans were buying. What suburban dream home of the 1950s would be complete without a full array of shiny new appliances and a flashy new automobile for the commute to the city? In the postwar economy, frugality, self-restraint, and doing without were no longer seen as virtues, and spending outsized portions of the household income on new consumer items was no longer seen as a vice. Rather, as Cohen notes, mass consumption was portrayed as a "civic responsibility" that would create new jobs and raise the standard of living for all Americans. Frustrated by more than fifteen years of depression and wartime shortages, Americans were happy to answer this call for consumption without guilt. New home purchases led the way in the postwar buying frenzy, but purchases to supply these homes with appliances—many of them electrical—followed close behind. For example, the jump in demand for electrical refrigerators was huge. In 1940, only 44 percent of Americans had mechanical refrigerators; by 1950, the number was 80 percent. Growth in new car sales was even greater, increasing by a factor of four between 1946 and 1955. All told, the national output of goods and services doubled between 1946 and 1956, with about two-thirds of this resulting from private consumer purchases.[48]

In most cases, the manufacture of all these new refrigerators and cars increased the demand for metals in the postwar years, including copper. The climb in domestic mining of copper did not precisely match the breakneck growth in the manufacture of consumer items for a variety of reasons. As the immediate postwar ads suggested, manufacturers could substitute aluminum for copper in some products or purchase recycled copper from scrap. Nonetheless, overall U.S. consumption of copper did grow by 45 percent between 1950 and 1974, while the output of copper from domestic mines and mills doubled from a postwar low of 797,000 tons per year in 1946 to 1,600,000 tons by 1970.[49] Further, an increasing share of this output was being used in consumer items, particularly houses, automobiles, radios, televisions, and other electrical products.[50]

In the 1950s, the Kennecott Copper Corporation, which acquired Jackling's Bingham mine in 1936, also published a series of clever ads linking copper and consumption. In a 1958 ad titled "Before You Put Appliances In, Put 'Skimpy Wiring' Out," a housewife sweeps the "Skimpy Wire" man—a stick figure with a light bulb head and thin copper wire limbs—out of her home just as a new chrome electric toaster oven is being delivered. The theme continues in "Skimpy Wiring can keep out the air conditioner you need!" where the evil symbol of inadequately thick copper wiring maliciously pulls on the power cord of a new air conditioner, thus threatening to deny the benefits of cool air to two cute kids perched in the home's picture window. Other Kennecott ads stressed the importance of the Bingham Pit copper that actually went into many popular new electronic devices of the time. In a 1957 offering, once again featuring a pair of attractive children, now raptly watching a television set, Kennecott insists that "in electronics . . . no substitute can do what copper does!" Copper and its alloys, it continues, "make possible more economical mass production of these modern marvels."[51]

By the middle of the 1950s, fears of a return to Depression-era levels of copper consumption had eased as demand remained reasonably robust, thanks to the rapid growth in the consumer economy as well as the renewed military need for copper created by the cold war. The copper advertisements reflected this improved economic climate with themes that shifted away from the merely utilitarian value of copper and toward its supposed role in achieving and protecting higher American ideals. One of the more intriguing examples is a 1950 ad illustrating the ringing of the Liberty Bell, that archetypal symbol of American patriotism that was made of the copper and tin alloy bronze. "Copper," the ad copy asserts, "helps freedom ring." Just as copper bells had once communicated the message of liberty in 1776, so in 1950 copper wire "carries to the world the news which helps to keep men free." Suggesting copper's role in the ideological battles of the emerging cold war, the ad notes that telephones, telegraphs, radios, and televisions all depended on copper to carry the " 'voice' of freedom." But this global ideological conflict was equally a struggle between competing economic systems: "For copper, and its alloys brass and bronze, play a vital role in enabling free men to make in abundance the products that keep our economy strong."[52]

In 1951, with the nation bogged down in a difficult war in Korea, the Anaconda embellished these themes with a series of ads comparing consumer and military uses of copper. One ad picturing a father and son happily working in their basement woodshop asked, "How many workshop motors to make a howitzer?" Another wondered rhetorically, "How many bathrooms in a jet plane's engine?" The ads went on to report that it took the copper from 210 workshop motors to build one 105 mm howitzer and that "it may take as much copper and brass to build a jet plane engine as there is in the faucets, piping, shower heads and all the other fittings of 3 bathrooms." On the one hand, these ads simply furthered the previous campaign to educate Americans about the consumer uses of copper and to associate the metal with wholesome family activities like father-son carpentry projects. In the midst of a war against Communist powers, however, the ads also suggested that the nation needed to limit the civilian use of copper in "order to meet our military requirements."[53] Implicitly, the ads suggested Americans must temporarily forego the joys of consumerism in order to meet the greater threat of global Communism. Once the world had been made safe for consumerism, Americans could again freely enjoy modern bathrooms and basement workshops filled with copper.

COPPER...

Helps freedom ring

It was the ringing voice of copper that on July 4th, 1776, proclaimed "LIBERTY THROUGHOUT ALL THE LAND UNTO ALL THE INHABITANTS THEREOF." Today, copper, in the form of gleaming strands of wire, carries to the world the news which helps to keep men free.

Were it not for copper, modern communication systems would be practically non-existent. For telephone and telegraph, radio and television can speed our word or image to its farthest destination as swiftly as light itself; but only because copper provides a pathway for the electricity that gives life to these communications.

More than the "voice" of freedom, copper is its strong helping hand, too. For copper, and its alloys brass and bronze, play a vital role in enabling free men to make in abundance the products that keep our economy strong.

Anaconda, foremost in copper, brass and bronze, is proud that the products of its mines and mills are helping to strengthen the cause of Freedom.

ANACONDA

First in Copper, Brass and Bronze

"Anaconda" is a registered trademark.   60355

Forty nine exact replicas of the famous Liberty Bell have been presented by the Copper Industry to the Treasury Department. It was Copper, the Voice of Liberty, that on May 15th announced the opening of the 1950 Independence Savings Bond Drive.

ACCORDING TO LEGEND, THE SIGNAL TO RING OUT THE NEWS THAT THE DECLARATION OF INDEPENDENCE HAD BEEN SIGNED WAS GIVEN BY A SMALL BOY WHO CLIMBED TO THE BELFRY OF THE STATE HOUSE IN PHILADELPHIA AND SHOUTED, "RING! RING!"

18. In this 1950 advertisement, the Anaconda moved beyond earlier publicity campaigns emphasizing the simple utility of copper to suggest the metal was also instrumental in fostering and protecting such grand ideals as American liberty, both past and present. In the emerging cold war world, the ad hints, copper would help "free men" create a robust economy that would soundly defeat that of their Communist enemies. *Author's collection.*

COPPER...*Most useful metal known to man*

How many workshop motors to make a howitzer?

Take the copper in one motor on these workshop tools and multiply by 210. That's how
much copper you will need to build one
105 mm. howitzer!

And it takes all the copper in your kitchen
—in your refrigerator, range, washer, mixer,
toaster—to make the ammunition that one
of our .50 caliber machine guns fires in less
than a minute!

These are but two examples of the huge

poundage of copper and its alloys, brass and
bronze, needed to meet our military requirements. There will be enough copper for the
weapons to defend America, but there cannot
be enough for all the less essential needs as
well. That is why it has been necessary to
limit the amount of copper for civilian use.

There is no real substitute for copper. No
other metal can match its unique combination of advantages. Copper is needed for its

strength and workability, its resistance to corrosion, immunity to rust, unusual ability to
conduct heat and electricity.

Today the copper industry must be more
concerned with howitzers than with workshop
motors. We of Anaconda are running every
resource—mines and refineries, brass mills
and wire mills—to fill every military require-
ment, and as many civilian needs as we
possibly can.

First in Copper, Brass and Bronze **ANACONDA**

19. While still reminding readers of the importance of copper to a happy domestic life, this advertisement from the early 1950s also suggests Americans
would have to do with less copper during the war in Korea. Once the world had
been made safe for consumerism, however, copper could again be used in Father's workshops instead of howitzers. *Author's collection.*

This theme of the cold war as a competition between two economic systems was reiterated a few years later when Isaac F. Marcosson, a writer of contracted corporate histories, published his celebratory story of the Anaconda. Here Marcosson argues that copper was essential to the realization of the "Machine Age" in the United States, noting that American per capita consumption of copper in 1925 was 16.24 pounds, nearly eight times the rate in the other nations of the "Free World." The "non-free" and presumably Soviet bloc nations, he pointedly stresses, consumed only a feeble 1.4 pounds per capita. "In these figures," Marcosson concludes, "you have an illuminating commentary on the American way of life to which the red metal is such a contributory factor."[54]

Thus, through a variety of venues, Anaconda and Kennecott suggested that their copper mines were not just the first source for all sorts of desirable consumer products, but that they also provided the foundations for American freedom itself—a freedom built on the industrial economics of capitalist mass production and consumption. Still, with the exception of the Marcosson book, in all of this publicity material the actual source of the copper—the landscape of extraction—remains implicit but invisible. This is why the second type of promotions the Anaconda began to produce in the mid-1950s is so significant. Ads like the "See America the Bountiful" series and the promotion of tourism in Butte and Bingham all served to redefine the meaning of the actual technological landscape of extraction, seeking to transform gigantic open-pit mines into populist symbols of the infinite economic growth and abundance that made possible the postwar "American way of life."

The Anaconda was not the first company to promote its open-pit copper mine as a tourist destination. Well before the 1950s, Jackling's huge Bingham Canyon mine had become one of Utah's major attractions. Even during the early stages of operation, John D. Rockefeller called the pit "the greatest industrial spectacle in the world."[55] And the mining engineer T. A. Rickard felt awed by the scale of all that industrial might: "It is a huge theater, in which the actors are 1800 men; but so big is the stage that they are hardly discernible at this distance. Ore-trains, like children's toys seen from afar, run along the levels and black steam shovels vomit puffs of smoke as they dig energetically into the precipitous face of the artificial cliff."[56]

Other commentators focused on the aesthetic and symbolic meaning of the pit. In 1943, one admirer of the pit penned a self-published tribute, *Thoughts and Meditations upon Beholding the Utah Copper Mine.* "The pit reflects a glistening greenish, bluish and often reddish hue of colors,"

Otto P. Chendron writes. Observers from some other cultures, like the mountain-worshipping Chinese, might consider the pit to be a ghastly mutilation. But the American observer, Chendron argues, appreciated the true significance of the pit for the very same reasons that American industrial culture had diverged from these other cultures. By using the pit to provide cheap copper for cheap electric power, American industrial might "puts the mechanical slaves into the home of the poorest, lights his lamps, sweeps his floor, runs with his messages, turns the dark night into working day." Meanwhile, the Chinese huddled in, presumably, dark, unswept shacks in the shadows of their sacred mountains.[57]

As the size of the pit grew in the postwar period, it attracted ever more visitors seeking to be awed by the sheer immensity of the mine. For a small fee, tourists could drive up the narrow canyon to an overlook where the signs and displays at a visitors' center encouraged them to view the pit in heroic terms. In 1965 alone, some two hundred thousand people visited the pit, where they were encouraged to view it as a graceful pairing of man and nature, reviving the old idea that the American exploitation of the continent has always somehow been a part of nature's master plan. As one architectural student insists in his thesis project on the site, the pit is "a living monument of sculptured beauty from earth forms carved from nature's mountains by ambitious men and their equipment. Although conceived by man, they seem to say that if a mountain must be moved, this would be nature's way of doing it."[58]

Perhaps because its excavation began in 1955, precisely at the point when the American "consumer's republic" was first being fully realized, the Anaconda's publicity for the Berkeley Pit surpassed that for the Bingham Pit in connecting the landscape of extraction to the consumer economy. The 1957 *Saturday Review* advertisement makes these connections explicit by emphasizing the pit's role in providing "the metals so essential to the nation's economic strength." Rather than linking copper ore to specific industrial and consumer products, as the earlier advertising campaigns had done, the 1957 ad stresses simply the sheer abundance made possible by open-pit mining, the "seemingly inexhaustible mineral wealth of . . . an area whose output increases year after year." The ad was also a departure from previous Anaconda publicity campaigns in that it encouraged readers to actually visit Butte itself, to come and "stand on a ledge 500 feet above the floor of the new Berkeley 'canyon,' " where they will gain "a new pride and appreciation of your America the Bountiful."[59]

If a reader responded to the ad and made the trip to Butte, the company stood ready to reinforce these messages with free educational pamphlets distributed at the Berkeley Pit viewing platform. Printed in the style of a daily newspaper, the *Anaconda Copper Trailsman* stressed the awesome size and technological sophistication of the pit operations. The first edition of the *Trailsman*, which appeared in 1956, offers the headline "Berkeley Open Pit Draws Thousands of Tourists: 17,500 Tons of Copper Ores per Day to Come from New Project on Butte Hill."[60] This and subsequent editions provided surprisingly detailed explanations of the technological operations of the pit. "The work goes on around the clock," one article explains, "blasting, loading, and moving more than 110,000 tons of ore every 24 hours." The pamphlets emphasized the sheer size of the rapidly growing pit, and they made effective use of photos and statistics to drive home the gargantuan stature of the trucks and shovels that moved these tons of ore so quickly. "Only the use of large, modern equipment makes such an operation possible," a 1965 article notes. "The pit requires nearly 250 vehicles, including 85 ore-hauling units, with capacity from 65 to 85 tons."[61]

Size, speed, and efficiency—these were indeed the keys to the successful operation of the Berkeley Pit, where ore as lean as one-half of 1 percent copper was mined. By providing a detailed technological explanation of how this industrial marvel of efficient mass extraction operated, the Anaconda was in part simply trying to satisfy the curiosity of the "sidewalk foreman," as one edition of the *Trailsman* put it. At a deeper level, though, this promotional material also suggests an attempt on the part of the Anaconda to assure Americans that the company's promise of "seemingly inexhaustible mineral wealth" was something more than just an idle boast. Though for obvious reasons the Anaconda never used the term, the pamphlets essentially suggest that the technology of mass destruction would make this modern-day metallic version of the miracle of the loaves and fishes a reality.

With the gigantic Berkeley Pit, the Anaconda had created a powerful symbol of the mineral foundations of economic prosperity. Though earlier underground mines at Butte had been equally vast and awesome with thousands of miles of tunnels carved from the rock, these operations had been largely invisible to the passing tourist. Aside from the head frames and surface plants that dotted the hill, there was little sign of the elaborate technological operation that lay beneath. By contrast, the Berkeley Pit and others like it laid bare the technological infrastructure of modernity itself. Here the public could experience a visceral example of the nation's productive industrial might, hear the roar of thirty-six-ton dump trucks, and feel even the earth itself shake from powerful blasts of high explosives. Here, the pit seemed to say, was the source of America's ever growing prosperity, the eternal wellsprings of the treasured American economic prosperity. As a later edition of the *Trailsman* notes, "Few people relate mining to their everyday living, yet we all are dependent upon it. So is the prosperity and strength of our nation. There's a slogan, 'From the Earth . . . a Better Life.'"[62] Or, as the Anaconda publicists might have phrased it in the 1950s, "From the Earth . . . the American Way of Life."

The Anaconda was able to tie the Berkeley Pit with the greater good of national prosperity at least in part because it was simply true. The Berkeley, Bingham, and other open-pit copper mines were critical foundations of the modern consumer society and economic growth, as were countless other iron, coal, and aluminum (bauxite) mines around the world. Though the copper mining industry was often plagued by overproduction, the cheap and readily available copper it produced nonetheless met an essential American (and increasingly global) economic need. However, this material prosperity was not achieved nearly so easily or painlessly as the Anaconda's promotional campaigns suggested. Left unmentioned in the company's Berkeley Pit publicity materials were the dead zones in Anaconda, the vanished cattle and forests, and the growing piles of toxic waste at the Opportunity tailings ponds. In this light, the Anaconda's boast that it could produce a nearly inexhaustible amount of copper from the pit was true, though only if the company was allowed to produce an equally inexhaustible amount of dangerous and stubbornly enduring waste. Likewise, the continuous physical growth of the pit had its costs. By the early 1960s, the company began tearing down or burying parts of the old urban core of Butte so that it could expand the pit. Given the company's earlier attempts to link copper mining with the rapid growth of postwar housing, these images of Anaconda employees tearing down or covering up perfectly habitable buildings in Butte were more than a little ironic.

The definitive falsification of the promise of painless infinite extraction and consumption came in 1982 when ARCO, Anaconda's corporate successor, shut down the Berkeley Pit. With that decision ARCO also chose to turn off the massive underground pumps that had kept the pit dry for more than a quarter of a century, and soon afterward toxic water began to seep into the rocky bowl. A year later, the pit won the dubious distinction of becoming part of the largest Superfund site in the nation, which also included the neighboring area around the Anaconda's Washoe smelter in Deer Lodge as well as the entire 120-mile course of the Clark Fork River

20. The city of Butte paid a steep price as the Anaconda attempted to make good on its ultimately illusory promise of nearly inexhaustible copper production. The company buried or leveled parts of the city in order to continue operations at the Berkeley Pit. Here one of the immense pit trucks (the tires are over eight feet high) dumps its load of overburden on the city's Holy Savior School. *Photo by Walter Hinick, from* Remembering Butte: Montana's Richest City *(Butte: Far-country Press, 2001).*

down to the Milltown dam outside of the city of Missoula. Although state and federal agencies and the responsible parties have since made considerable progress in the cleanup, the challenges to permanently remediating the site remain daunting. The water in the Berkeley Pit is a highly acidic toxic soup that carries 15 times the acceptable EPA level of arsenic for drinking water, 430 times the acceptable level of cadmium, and 7 times the lead. If nothing is done, scientists note, the rising tide of toxic groundwater will eventually bleed into the nearby Silver Bow Creek and thence into the Clark Fork River.[63] The consequences of such an event for aquatic life in the Clark Fork watershed would be, as one scientific report put it, "catastrophic in both nature and extent."[64] Further contamination of the already poisoned Butte groundwater would occur even sooner, potentially flooding the basements of thousands of the town's residents who live on "the Flat," an area below the Butte hill that had been a wetland before the Anaconda pumps lowered the water table by thousands of feet.[65]

Such an outcome is almost unthinkable, as it would require the mass evacuation of as many as half of Butte's thirty-three thousand residents. Currently, ARCO is using the only economically feasible treatment technology available to treat the water: adding lime to neutralize the acidity, which allows the metals to precipitate out into five hundred to a thousand tons of worthless hazardous sludge every day.[66] Research is ongoing by ARCO, Montana Tech, the EPA, and other state and federal interests to develop technologies that could profitably extract the valuable metallic contents from the water, but success has been elusive.[67] Absent such a method, the toxic sludge is currently transported over to the Deer Lodge Valley, where it is added to the already immense piles of waste left behind in the Washoe plant's Opportunity tailings ponds. Regardless, whatever the strategy adopted to clean the Berkeley Pit water, treatment will have to continue for centuries to come before the daily flow of 2.5 million gallons of groundwater might leach all of the poisons from the ten thousand miles of underground passages left by a century of mining.

The consequences of the mass destruction operations at Jackling's Bingham mine have been somewhat different, for two main reasons. First, the mine continues to be operated by its current owner, Rio Tinto. Second, the Bingham Pit does not intersect groundwater flows as the Berkeley Pit does, greatly reducing the problem of acid mine drainage. However, the Bingham Pit does face many of the same problems with vast amounts of dangerously toxic heavy metal tailings. In 1994, the EPA identified the (then Kennecott-owned) Bingham mining complex as one of the worst sources of toxic waste in the nation. Indeed, the EPA considered the pollution problem at Bingham so serious that it threatened to place the entire complex on the National Priorities List (NPL), which would allow the federal government to require the company to take steps toward environmental remediation using the Superfund process. Such a move would have been unprecedented. The EPA had previously designated many inactive mines for Superfund remediation, like the Berkeley Pit and Anaconda smelter sites, as well as some parts of active mines. But this was the first time the agency raised the possibility that it might be necessary to take over all fifty-seven square miles of an active mining operation, encompassing the pit, concentrating mills, sulfuric acid plants, slurry lines, tailings heap, slag piles, smelter, and refinery.[68]

The EPA investigations of the Bingham complex revealed multiple sources of toxic contamination. Sulfates had drained into the groundwater, poisoning wells over an area of more than seventy-seven square miles. In other areas, EPA researchers discovered concentrations of lead and arsenic in soil and groundwater far above levels recommended for human exposure.[69] Pollution is also escaping from the concentrating and smelting complex along the shores of the Great Salt Lake. At the huge Magna concentrator site that Jackling and the Guggenheims began using in 1905, the EPA found high levels of lead and arsenic in the immense sludge ponds, tailings, and slag piles.[70]

The company avoided Superfund listing by agreeing to begin a voluntary cleanup effort.[71] By 1997, Kennecott had spent over $150 million on moving and capping twenty-five million tons of contaminated material and installing liners and leak detectors beneath mine wastewater collection channels.[72] Yet these efforts barely began to remediate the extensive environmental damage at the Bingham complex, and they did little to solve the long-term problems posed by the two billion tons of tailings perched on the shores of the Great Salt Lake.[73] Moreover, despite the company having recently opened new ultra-efficient concentrating and smelting facilities incorporating advanced pollution controls, in 1996 the EPA placed the Utah Copper operations on its Toxic Release Inventory (TRI) as the fourteenth largest producer of pollutants in the United States. Utah Copper had released 9.6 million pounds of pollutants into the environment that year.[74] Currently, the government does not require hard-rock mine operators (versus stand-alone smelting or refining operators) to divulge the amount of waste they produce from concentrating mills and other mine operations for the TRI. However, in 1989 the Utah Copper Corporation mistakenly believed it was required to report its entire waste production and disclosed that it had produced over 158 million pounds of waste in 1987. In contrast, the Magnesium Corporation of America, which was at the top of the TRI listing, produced only 55.7 million pounds of waste in 1996.[75]

Whether at Bingham, Berkeley, or any of the hundreds of other open-pit mining operations around the world, the bill for a century of cheap copper will continue to come due for many years. Today, as was the case nearly fifty years ago, the Berkeley Pit continues to be a stop for tourists as they travel to Yellowstone and Glacier parks, although no one runs full-page ads in popular magazines to encourage their visits. People seem to make the trip anyway, coming to gaze down at the poisonous dark waters of the slowly growing "Berkeley Lake." Some of them may reasonably conclude that such environmental degradation is the price the nation has paid for a reckless pursuit of unending economic growth and expansion. In this light, we can see that "America the beautiful" was often a victim of "America the bountiful." It is equally important to realize, though, that Anaconda and Kennecott's basic goal of illuminating the links between mass mining and mass consumerism is no less accurate and relevant today than it was in the 1950s, though we can now better understand the negative as well as the positive consequences of these links. If anything, in the early twenty-first-century Americans seem to be even more addicted to the political, social, and economic blessings of mass consumerism than they were half a century ago. Further, hundreds of millions of people in China, India, Brazil, and other nations around the globe are now eagerly poised to begin replicating American levels of consumption. While efficient use, recycling, and improved mining techniques may soften the blow, the raw materials for a billion new cars and a billion new refrigerators will almost certainly require more mass destruction mines and more environmental degradation. Thus, while the Berkeley Pit today stands quiet, we can be sure that somewhere around the world new landscapes of extraction promising seemingly unending bounty are being created. The culture and technologies of mass destruction seem destined to expand even further before they are likely to undergo any fundamental changes.

ALL THE WORLD'S A MINE

Only five years after Jackling had demonstrated that low-grade porphyry copper ores could be profitably mined with his system of mass destruction, American mining engineers had already slated a dozen similar ore bodies for similar treatment. The three major copper mining companies—Anaconda, Phelps Dodge, and Utah Copper/Kennecott—soon began digging open pits all around the west, including the Ray, Miami, Sacramento Hill, and Inspiration pits in Arizona; the Chino Copper Company pit in New Mexico; and eventually the Berkeley Pit in Montana. Wherever mining geologists could discover suitably large and shallow low-grade copper deposits, the efficiency of the open pit replaced the expensive and dangerous process of underground mining. Labor productivity soared while costs plummeted.[76] In 1911, the average copper mine required 110,000,000 hours of human work to make 557,000 tons of copper; by 1949, it took a third less labor to make a third more copper. In 1911, underground mining techniques extracted 0.27 tons of copper for each hour of human labor; by

1949, that number had risen to 2.20 tons, nearly a tenfold increase, in large part due to the widespread adoption of Jackling's mass destruction mining techniques.[77]

This astounding increase in labor productivity was fundamentally possible because open-pit mines allowed Jackling and others to use machines that substituted hydrocarbons for human energy. Much of the economic savings of mass destruction mining came from diminished labor costs made possible by the steam (and later electric, diesel, and gas) shovels and engines by means of which one human moving a lever could do the work that had previously demanded dozens of men. To be sure, many mining men, including Jackling, favored such methods in part because they detested unions, strikes, and all the other ways they believed human workers interfered with their supposed right to maximize their own and corporate profits. But it would be a mistake to overemphasize class and labor issues without also recognizing the equally important role of increased technological efficiency and energy use. Again, a few statistics make the scale of the transformation clear. In 1880, the American copper mining industry had about 13,500 horsepower of machinery. By 1939, the figure had jumped to 750,000 horsepower, a fiftyfold increase, primarily because of the growing use of power shovels, ore trains and trucks, and steam-powered mill and smelter machinery. To avoid capturing merely the effects of increased mining, consider the shift in horsepower consumed per worker. In 1880, copper mines had 2.2 horsepower available per worker. In 1939, the number was nearly twelve times greater: 26 horsepower per worker. Strikingly, these figures do not even include the rapid growth in the use of electrical power (much of it hydropower, but a growing amount from coal) in copper mining, which was virtually nonexistent in 1900 but had reached 32 electric horsepower per worker by 1939.[78]

The conclusion is obvious: Jackling's mass destruction mining system achieved its huge advances in speed, scale, and productivity through the profligate use of coal, oil, and hydroelectric energy resources. Mass destruction mining ultimately only made sense because these energy sources were relatively cheap and available throughout much of the twentieth century. Open pits permitted the copper industry to join what the historian Alfred Chandler identifies as a national industrial drive toward the increased use of "capital-intensive, energy-consuming, continuous or large-batch production technology to produce for mass markets."[79] In essence, Jackling created a mass extraction and processing system for minerals similar to those first used in industries processing liquids and semiliquids like petroleum. Indeed, much of the Bingham system—from the initial ore breaking with dynamite all the way down to the fine grinding process for concentration—was designed to make rock into a powder that could be mixed with water and handled like a liquid. Mass destruction mining stuffed huge portions of the natural world into its grinding machinery and spewed out a small amount of concentrated copper along with immense amounts of waste. Cheap coal, oil, and electricity made such mass destruction not only feasible but even logical and seemingly efficient.

The success of Jackling's open-pit mining system also helped to shift the manner in which industrialized society viewed natural resources. Mass destruction strengthened the already powerful tendency among Western industrialized nations to see the natural world as an abstracted economic commodity, a mere product waiting to be mined and processed for the market.[80] As one copper analyst noted, the increased use of machinery and energy in mining meant that "depletion . . . occurs only in a relative sense, in that grade declines and natural conditions become more difficult, but absolute exhaustion rarely takes place."[81] By 1944, more than half of the nation's copper came from deposits discovered as early as 1900 but considered valueless.[82] Given enough energy and the proper technology, there seemed to be no reason to think that the extension of what constituted economically valuable ore would stop. Many mining engineers had thought Jackling was foolish to attempt mining on a 2 percent ore deposit in 1905. Yet by 1960 the Utah Copper Company was profitably exploiting ore that averaged only four-tenths of 1 percent—a mere eight pounds of copper per ton of ore.[83] Taken to its logical extreme, this view of natural resources suggested that almost any area of the planet was a potential commodity waiting to be mined. In the 1980s, a group of Japanese investors even experimented with extracting the minuscule amounts of gold dissolved in seawater, an idea suggested by some metallurgists as early as 1872. With only one gram of gold in a hundred million tons of water, seawater mining has yet to prove economically practical, but others will likely return to the proposition if future energy supplies prove adequate to the task.[84]

Barring the development of cheap and powerful new energy sources anytime in the near future (and assuming continued limits on the development of nuclear power), however, the century-long trend toward the exploitation of ever leaner ores must eventually encounter the limits of available power. The earth has only a finite supply of coal, oil, and dammable rivers, and the problem of global warming makes increased use of hydrocarbons dangerous. In fact, the environmental costs of Jackling's mass destruction may well surpass its benefits before reaching these energy limits. The combined environmental costs of hundreds of open-pit copper mines (as well as similar pits for gold, iron, bauxite, and other materials) created around the globe are only beginning to be assessed and understood, but like those at Butte and Anaconda, they are often severe. Taking into account all forms of open-pit mining (including quarrying for building materials like gravel and sand and strip-mining for coal), by 1980 giant shovels, trains, and trucks moved more earth on the planet than did the forces of natural erosion. Mining had literally become equivalent to a force of nature. In 1990, American open-pit mining operations alone moved an astounding four billion tons of rock per year, while the figure for the entire globe may have been as much as twenty billion tons.[85]

Just how far mass destruction technology may yet spread is unclear. In 1941, the mining engineer E. D. Gardner argued that because gold and silver ore deposits were relatively small, these precious minerals "seldom lend themselves to large-scale exploitation by open-cut methods."[86] But by 1963, the geology and economics of mining had changed, and some 90 percent of all the metal produced in the United States, including the precious metals, originated in open pits. During the 1950s and 1960s, several American geologists identified a geological structure known as the Carlin Trend, a forty-mile-long region in Nevada filled with so-called invisible gold, ultra-low-grade ore in which microscopically small specks of gold are uniformly disseminated in masses of rock in a manner roughly analogous to Jackling's porphyry copper deposits. Combining Jackling's energy- and capital-intensive mass destruction technology with a new metallurgical process called cyanide heap-leaching, U.S. gold production at the Carlin Trend and similar deposits increased tenfold during the 1980s.[87] With gold prices high, mining companies rushed to develop open-pit, cyanide heap-leaching gold mines all around the western United States.

The microscopic amounts of gold mined today make Jackling's 2 percent Bingham copper ore look rich. The highest-grade deposits in the Carlin Trend average about 0.20 ounces of gold per ton of rock, or less than 7/10,000 of 1 percent.[88] In a typical open-pit gold mine the operator thus mines almost three tons of ore to produce enough gold to make one small wedding band, while the remainder is left behind as huge volumes of waste. The resulting environmental damage and accompanying social dislocation can be particularly severe in foreign countries that lack stringent regulations. An open-pit gold mine operation in Irian Jaya, Indonesia, opened in the late 1980s, generates about 120,000 tons of toxic waste every day and is projected to generate 2.8 billion tons during its productive life.[89] Hundreds of similar open-pit mines were developed in Latin America, Africa, and Asia during the 1990s. Record prices for many minerals in the first decade of the twenty-first century have only accelerated the trend toward increased exploration and development, while also sparking the resurgence of mining in some long-abandoned old districts.[90] The specific mining and processing techniques adopted at these mines varied widely depending on the types of minerals, geology and topography of the site, regulatory environment, and other factors. Nonetheless, all of these hard-rock open-pit mining operations in one way or another drew on elements of the high-speed mass destruction mining technology Jackling pioneered at the Bingham Pit.

The technologies and ideologies of mass destruction have also become increasingly important in coal mining, though the geology of most coal deposits is very different from that of a typical hard-rock mine. The strip mining of coal actually preceded Jackling's development of the Bingham Pit. As early as 1866, miners used simple horse-drawn scrapers to remove the earth from a shallow coal seam in Illinois, and steam shovels were adopted for stripping by the 1870s.[91] Once exposed, the shallow and often immensely large and rich coal seams presented few of the challenges Jackling faced in extracting and processing very low-grade porphyry deposits with tiny amounts of copper dispersed throughout large masses of worthless rock. Still, as with Jackling's system, coal strip mining was far more efficient than traditional underground coal mining, in part because steam-, gas-, and electric-powered shovels took the place of human workers. By the end of the twentieth century, strip mine workers produced about four times as much coal for every hour of work than did their underground colleagues.[92]

The more recent development of so-called mountaintop removal coal mining techniques bears a closer resemblance to Jackling's system of mass destruction. Developed in the 1980s and 1990s in the mountainous Appalachian coalfields of southern West Virginia and eastern Kentucky, mountaintop removal mining offered a highly efficient but extraordinarily destructive alternative to previous underground mining methods. In contrast to the typical operations of the past where miners stripped the overburden from shallow coal deposits, these deposits were five hundred or even a thousand feet below the surface, and multiple layers of coal were interspersed with layers of rock and soil. To mine such deposits underground had required excavating many levels of tunnels on each seam to avoid moving the masses of waste rock and dirt between. As the name suggests, in mountaintop removal mining the operators instead use high explosives and huge electric-powered shovels and draglines to level entire mountains. The draglines used in these mines make even the big shovels at Bingham look puny. Some have buckets big enough to hold 110 cubic yards of earth, roughly equivalent to scooping up twenty-six Ford Escorts. House-sized dump trucks, similar to those used at Bingham and Berkeley, move the vast volumes of waste rock to dump sites in narrow valleys nearby. Some of these "valley fills" are as much as a thousand feet wide, five hundred feet deep, and a mile in length. Many are deposited on top of mountain creeks, whose water carries silt, acidic drainage, and heavy metals like iron manganese downstream and into groundwater systems.[93] One recent estimate suggests that in West Virginia alone, the mines have leveled about four hundred square miles of mountains and ridges and buried a thousand miles of streams beneath debris.[94]

The creation of these new dead zones has sparked local and national movements to halt further mountaintop removal operations, though the protests have met with only mixed success. As with earlier forms of mass destruction mining, the operators argue that the technique is the only feasible method of profitably mining these types of coal deposits. Ironically, the companies can also make a legitimate claim that the high-energy, low-sulfur coal they are removing is better for the environment than the dirtier coal mined in many other regions.[95] Indeed, the demand for this low-sulfur coal to generate electricity skyrocketed after Congress passed the 1990 Clean Air Act.[96] As one manager of the big machines used in the mines argues, "Coal keeps the lights on."[97] Area inhabitants have also divided over the mines. In a state like West Virginia, where the biggest private employer is Wal-Mart and 16 percent of the population lives below the poverty line, the promise of relatively high-wage blue-collar jobs offered by the mining companies is enticing, if somewhat illusory. These highly mechanized modern mines employ far fewer (and now generally non-union) workers than did the underground mines of the past. During the previous ten years, coal mine output in the state increased by 32 percent, but the employment of coal miners dropped 29 percent.[98] Like Jackling's open pits, mountaintop removal mining achieves increased speed and efficiency by substituting powerful, crude, and energy-hungry machines for skilled human workers and transferring many operational costs to the environment.

The basic principles of mass destruction also emerged in extractive industries that might at first seem categorically different from copper or coal mining. One of the most striking examples was the North American timber industry's rapid adoption of the tree harvesting technique of clearcutting. Prior to the middle of the twentieth century, many timber operations around the nation still used selective harvesting techniques that depended heavily on skilled human labor and the use of draft animals for initial hauling. Using simple axes and crosscut saws, loggers felled only mature specimens of the desired species, leaving behind a still functioning and largely intact and diverse forest ecosystem that would continue to grow for future harvests. The loggers used horses or other draft animals to drag the felled trees to the yard, where they could be floated downstream on rivers and loaded onto trains or, by the post–World War II era, trucks.[99] To be sure, where the environment provided roughly uniform stands of mature trees that could be easily harvested at once, even these simple logging tools were sometimes used for clearcutting operations. However, because the task of felling and moving logs was so labor-intensive and time-consuming, loggers generally had an incentive to cut only the most profitable mature trees, much as skilled underground miners strove to extract only the richest ore.

Beginning in the early twentieth century, new technologies and ideologies began to undermine traditional selective harvest techniques. Many larger timber companies adopted so-called steam donkeys, mechanical engines designed to replace the living donkeys, horses, or other draft animals (as well as their teamsters) for hauling felled logs to the timber yards. Somewhat analogous to a steam-powered mine hoist, steam donkeys were basically big winches that used long manila or wire ropes to drag the felled logs through the forest to the yard. In subsequent decades, creative logging engineers further improved steam-powered hauling with overhead skidders that used elevated wire cables to lift one end of the log off the ground, making it much easier to drag rapidly over rough forest ground. Other steam- and later gas- and diesel-powered machinery for moving and loading the logs further increased the speed of timber extraction while decreasing the number of workers.[100] However, the actual extraction of the tree—felling by loggers using hand axes and saws—remained a labor-intensive bottleneck until after World War II. A viable solution only emerged when improved air-cooled engines and lightweight metals like aluminum made highly portable and maneuverable gasoline-powered chainsaws possible. By the late 1950s, chainsaws had essentially replaced hand tools throughout the timber industry for the simple reason that they allowed loggers to cut trees down as much as a thousand times faster as with axes and handsaws.[101]

The parallels between mechanized logging and Jackling's mass destruction mining system are apparent. Chainsaws sped the extraction of timber much as dynamite did for copper ore. Steam donkeys and overhead skidders sped the transport of the felled timber much as steam shovels and locomotives did for copper ore. The resulting increase in productivity was similarly impressive: in some regions timber output per hour of labor doubled between 1950 and 1968.[102] Unfortunately, high-speed energy-intensive timber harvesting also strongly encouraged logging companies to abandon selective logging in favor of clearcutting. To maximize the speed and efficiency of powered overhead skidders, timber companies preferred to remove all the trees in a cut so that workers could rapidly shift the overhead lines to new areas and nothing would impede the movement of the logs. Thanks to the speed of chainsaws, loggers could now quickly level vast acres of forests. Any small or valueless trees that escaped the chainsaws were often knocked flat by the big suspended logs as they were dragged out by the powerful skidder engines. The results were often ecologically catastrophic, leaving behind yet another type of dead zone that in some ecosystems might not be successfully reforested for many decades to come.[103] As one opponent of the clearcutting of the forests in the Bitterroot Mountains of Montana (not far to the west of Butte) suggested, this "isn't logging, it's mining."[104]

Like open-pit mining, clear-cutting of forests found ideological and institutional support within a broader culture of mass destruction. Richard A. Rajala, one of the leading historians of the timber industry in the Pacific Northwest, argues that the U.S. Forest Service accepted clearcutting as a "technological necessity" in order to meet the exploding postwar demand for timber products for housing, paper, and other consumer goods. Forest Service officials and other experts also justified the practice on a flawed scientific theory that clearcutting would maximize the regeneration of desirable Douglas fir stands.[105] Even the cold war struggle against Communism played a role, just as it did with the Anaconda's advertisements celebrating the Berkeley Pit as a foundation of American freedom. In a 1952 report by the President's Materials Policy Commission, the authors called for the maximum production of lumber and pulpwood because the struggle against "a new Dark Age" of Communism "must be supported by an ample materials base."[106] Likewise, as the historian Paul Hirt notes, during the fifties and sixties many American forest experts embraced a "conspiracy of optimism," which found broader resonance in the popular culture of the day. "In a sense," Hirt argues, "Americans became self-delusive in their enthusiasm for unending economic growth and techno-logical manipulation of natural systems in pursuit of wealth and national power."[107]

Just as mass destruction mining promised infinite supplies of copper, so too did mass destruction forestry promise infinite supplies of timber and cellulose pulp. The same was also true of some parts of the global fishing industry at midcentury. In 1955, two academics, one of them the director of the well-respected Marine and Fisheries Engineering Research Institute at Woods Hole, Massachusetts, published a book with the singularly appropriate title *The Inexhaustible Sea*. Noting the rapid growth in global population, the authors argue that humans must learn to make far more efficient use of the ocean's immense stocks of fish. Only then, they conclude, will we come to "learn that in its bounty the sea is inexhaustible."[108]

Such a claim might well have struck some as absurd, given that by 1955 global fisheries had already experienced frequent and severe localized declines from overfishing. The fishing industry, however, had repeatedly dealt with earlier population collapses by adopting new technologies that improved extractive efficiencies or opened up the exploitation of new fish populations around the globe. Many believed the extraordinary science and technology of the modern postwar world would continue to overcome any natural limits. For several decades, their confidence seemed justified, as the size of the global fishery catch increased steadily into the 1980s. The statistics, however, were misleading. Fishermen maintained these big harvests only by improving the fishing power of their ships—extracting increasingly "low-grade" fish concentrations much as miners extracted low-grade copper ore. Sharp increases in the extractive power of fishing boats first began in the second half of the nineteenth century with the development of steam-powered trawlers. The boats took their name from "beam trawls," a type of bag net in which the top of the mouth is held open by a wooden or steel horizontal beam while the weighted bottom is dragged over the ocean floor. Beam trawls scoop up everything in their path: vast numbers of fish, to be sure, but also sea fans, urchins, sea cucumbers, hagfish, and many other commercially worthless specimens. After fishermen winch the net on board, they pick out the handful of desired species and throw the remaining "by-catch"—most of it now dead or dying—back into the sea.[109]

Even during the age of sail, bottom trawling was both amazingly productive and astonishingly destructive compared to more selective traditional fishing methods like traps, hooks, and conventional nets. Still, sail trawlers were limited by the winds and tides, and they could not venture into areas where the seabed was rough and the trawl's ground rope might snag and break. Beginning in the 1870s, the use of steam-powered trawlers overcame these limitations. Steamships could operate with little concern about tides and weather while also towing far larger and stronger nets. By the early twentieth century, typical beam trawls were fifty feet wide. To access areas with rough bottoms, fishermen began wrapping the ground rope in chains so that it knocked over, crushed, or obliterated any obstacles in its way. Without the aid of steam-powered winches and wire ropes, it would have been nearly impossible to haul up these big heavy nets and their immense and indiscriminate mix of life from below. Steam winches were also far faster than earlier human-powered hand cranks, slashing the time it took to raise the trawl from fifty minutes to as little as fifteen. As a result, one fisheries expert estimated that the steam trawlers had about four times more fish-catching power than sail trawlers.[110]

The consequences were rapid and devastating. In the words of one marine scientist, by the late nineteenth century the steam trawls had already "caused the greatest transformation of marine habitats ever seen," wiping out immense areas of teeming marine life and leaving behind nearly sterile expanses of gravel, sand, and mud. It was, as one bait-and-hook fisherman suggested at the time, like operating a "gold mine."[111] The analogy to mining was, in retrospect, more accurate than he realized. Much like Jackling's big steam shovels, steam trawlers made rapid but nonselective extraction of a natural resource technologically and economically feasible. Likewise, just as the size and power of steam shovels grew steadily throughout the twentieth century, so did those of the steam (and later diesel) trawlers. After World War II, nations around the globe began building huge "factory trawlers," which incorporated fish freezing and processing facilities on board. Capable of staying at sea for months at a time, these giant ships could exploit fisheries far from their home ports. The first began trawling the Grand Banks of the North Atlantic in 1956—just one year after the Anaconda opened its Berkeley Pit.[112] Eventually, the powerful ships were hauling trawl nets with mouths nearly as wide as a football field and four stories high. Tellingly, the author William Warner turned to an analogy from timber clearcutting to describe the effects: "Try to imagine a mobile and completely self-contained timber-cutting machine that could smash through the roughest trails of the forest, cut down trees, mill them, and deliver consumer-ready lumber in half the time of normal logging and milling operations. This was exactly what factory trawlers did—this was exactly their effect on fish—in the forests of the deep."[113]

Trawling is not the only nonselective modern fishing method, though it has been one of the most destructive. This form of mass destruction fishing is also, as was the case in mining and logging, highly productive. During the twentieth century, the ocean's fish fed more human beings than ever before. In 1900, the global marine fish catch was about two million metric tons. By 1996, it was seventy-four million metric tons—enough to feed the estimated one billion people on the planet who depended on fish as their main source of animal protein. Thanks in no small part to the nonselective mass destruction techniques of trawling, fishermen in the 1980s could meet this demand by catching as much in two years as their predecessors had caught over the course of the entire nineteenth century. Again, such efficiencies came at a price. By the 1990s, the inadvertent by-catch of economically worthless marine life discarded by fishermen was about a third of the total catch—more than forty-seven million metric tons during the first six years of the 1990s. Further, precisely because they catch and throw away so much marine life, modern mass destruction fishing techniques have greatly increased the speed of harvesting the desired fish species. As a result, fisheries around the globe have seen repeated collapses in the decades following World War II.[114] Unless the world's nations take drastic cooperative measures soon, some experts warn, further declines may lead to severe global fish shortages by midcentury.[115]

Coal companies, loggers, and fishermen did not, of course, develop mountaintop removal mining, clearcutting, and factory trawling by consciously adapting Daniel Jackling's mining methods to their industries. Rather, what these striking parallels between very different extractive industries suggest is the centrality of a deeper ethos of mass destruction at the heart of the modern world, a common source of closely related technologies and ideologies that has fed the growth of many modern enterprises. Others have recognized these patterns, but most have missed their most essential features. Attempting to capture the essence of the trend, they have resorted to suggestive but ultimately misleading terms like industrial fishing and logging, industrialized nature, Fordism in nature, factories in nature, mass production in nature, and others. All of these concepts have some explanatory power, to be sure, but they also all fail to recognize or capture the essential defining trait shared by modern extractive mining, logging, and fishing. This was not merely the adoption of industrial power sources and machines, though the use of devices powered by coal and oil was essential. Nor was it just a matter of applying mass production factory techniques to speed the processing of natural materials, though the ideas of Ford and Taylor had their place. Nor was the central goal the de-skilling of labor, though the reduction in labor costs was critical. Rather, the basic principle shared by all of these modern extractive industries was this: the use of crude but powerful hydrocarbon-fueled machines to rapidly and efficiently chew up entire swaths of the world and to subsequently extract and keep only that small portion of it that was valued as a natural resource. Whether the immense area destroyed was a mountain, a forest, or a fishery, the process can only appropriately be termed mass destruction. This was the lever that moved much of the modern world.

In this broader sense, Daniel Jackling may have been the father of mass destruction mining, but he was also only one of many who perfected the wider technology of modern mass destruction. Still, Jackling's Bingham Pit offers a fitting archetype for the rest.[116] Even when applied to theoretically renewable resources like trees and fish, mass destruction techniques were so powerful and unsustainable that they often really did constitute a type of mining. The extraction of fiber and flesh happened so quickly and at such an immense scale that the targeted organisms and ecosystems had little chance to regenerate naturally. Likewise, if mass destruction techniques could promise to provide a nearly infinite supply of minerals from an inherently limited resource, then surely it is no surprise that some believed they could do the same for renewable resources like trees and fish. Thus did mass destruction promise a modern technological realization of both the ancient biblical miracle of the loaves and fishes and the medieval alchemist's dream of transmuting lead into gold: from scarcity would come abundance, from waste would come treasure.

How fitting, then, that in awarding him the 1933 Fritz Medal for outstanding achievement in mining and metallurgy, Jackling's fellow engineers celebrated his method for the winning of "mere traces of metal in the rock mass" through which the "superskill of man can transmute rock into ore." So it was, the award citation concluded, "that Nature's unconditional surrender to man is best exemplified."[117]

Ironically, only twelve years later the Japanese failure to heed the American demand for an immediate and "unconditional surrender" would be met by the atomic annihilation of Hiroshima and Nagasaki. Whether the enemy was nature or other human beings, the technology and culture of mass destruction seemed to demand nothing less than total submission.

EPILOGUE

From New Delhi
to the New West

Here, we live in darkness.

—Villager in Chakai Haat, India

On a Friday night in Gurgaon, a booming suburb of India's capital of New Delhi, honking cars crowd the streets and commuter traffic slows to a crawl. Above the busy streets and sidewalks, the brightly lit windows of new air-conditioned malls and office buildings gleam in the gathering darkness, some with twenty-foot-high illuminated advertisements for pricey handbags, clothes, and electronics. One of the shiny modern office buildings is the Gurgaon base of Ericsson, a Swedish telecommunications company that has pioneered the adaption of existing copper wire networks for broadband information transmission. In India's increasingly high-tech information economy, old-tech copper wires will still play an important role.[1]

Beyond the glittering affluence of central Gurgaon and other urban enclaves, however, the need for copper and the power it carries remains much more basic. In the impoverished shantytowns surrounding the city, the electricity still frequently fails, while outside the urban centers almost half of India's 1.3 billion people have no access to the electricity grid at all. In many small rural villages, like Chakai Haat near India's northern border with Nepal, bicycles and horse-drawn carts are more common than automobiles. A diesel generator provides a few brief hours of electric light at night, but after the engines shut down the villagers rely on kerosene lamps and battery-powered flashlights. Most burn gathered sticks and dried animal dung to fuel their cookstoves. As one villager who had previously enjoyed the more modern amenities offered in the city of Mumbai said, "There, we live in light. Here, we live in darkness."[2]

India's government has promised to extend the electrical power grid into more rural areas during the next five years and to increase the supply and reliability of power nationwide. Though it may be some time yet before Chakai Haat has the bright lights, air conditioners, and busy traffic of Gurgaon or Mumbai, the swift economic growth and modernization of India suggests they are coming. The prospect alarms many environmentalists and policy makers who worry about the effect of India's rapidly growing energy consumption on global warming. The American per capita carbon footprint is currently sixteen times that of the average Indian. If 1.3 billion Indians were all to match the extravagant American levels of energy use and carbon emissions, efforts to limit or decrease global greenhouse gases might well be futile. India's government leaders, however, understandably resist pressure from industrialized or postindustrialized nations like the United States to limit their carbon emissions. With half of India still blanketed in darkness every night, it is difficult to argue that the Indians should not have the right to use more energy rather than less.

Given the growing worries over global warming, concerns about the breakneck pace of modernization in India, China, Brazil, and other countries have focused on their increasing consumption of oil, coal, and other hydrocarbon fuels. However, as the conditions in the village of Chakai Haat suggest, extending electric power to rural areas and providing more Indians with air conditioners and cars will also entail increased consumption of mineral resources like copper. India is unlikely to mirror the historic American patterns of copper use exactly. Today, for example, most new electric power lines are made from aluminum, while cell phone technology and fiber optics can eliminate the need for copper-based telephone lines. Nonetheless, copper continues to be the metal of choice for interior wiring, plumbing, electric motors, and many popular consumer electronics like air conditioners, refrigerators, and televisions. Likewise, a typical three-thousand-pound modern automobile now contains more than forty pounds of copper. Indeed, given that copper is essential to so many modern products and services, its consumption often mirrors broader trends in economic growth and industrialization. In recent years, China and India have both rapidly increased their use of copper. In 1974, China's copper consumption was a modest 0.71 pounds per person; by 2007, it had increased more than tenfold to 8.1 pounds. During the same period, India's copper consumption jumped fivefold to 0.86 pounds per person.[3]

Much of this increased demand for copper has occurred during the last five years, when the supply of copper on the global market was heading in the opposite direction. As a result, copper prices have reached and surpassed previous historic highs. In January 2003, a pound of copper sold for about $.70, but by June of 2008 the price was over $3.80. The subsequent global economic decline rapidly squelched demand and drove prices below the two-dollar mark by year's end. But assuming that a recovery will eventually follow even a deep recession, it is reasonable to believe that the world's hunger for copper will soon resume its historic climb upward. Analysts debate whether these trends will presage an impending global copper famine in the decades to come. In May 2006, USGS resource geologist Donald Menzie testified on the subject during a congressional hearing on energy and mineral requirements. Menzie argued that while continued rapid increases in copper consumption in the developing world may produce temporary shortages, global copper deposits would prove adequate to meet demands through the year 2020. However, Menzie admitted that (assuming recycling of existing copper stocks continues to supply about 10 percent) to meet future demands mining companies will need to add about 1.1 billion tons of copper to the known reserves. Most of this would likely come from yet-to-be-discovered big low-grade copper deposits where for each new ton of copper, mining and processing would generate about 350 tons of waste rock and 147 tons of tailings. Barring some major technological breakthrough, Menzie estimated, meeting the soaring world demand for copper between 2000 and 2020 would thus generate about 130 billion tons of waste rock and 56 billion tons of tailings—a total of about 186 billion tons of waste with varying levels of toxicity.[4]

To put this number in perspective, consider that by one government estimate the entire noncoal mining industry of the United States currently produces between one and two billion tons of waste annually—that includes the waste from the Bingham Pit, the big Morenci and Ray pits in Arizona, the El Chino Pit in New Mexico, the vast open-pit gold mines of Nevada's Carlin Trend, and many others.[5] Thus, while it may be technologically feasible to produce enough copper to feed global demand through 2020, it will require adding a staggering amount of pollution to the planet: roughly the equivalent of four to nine times the total waste currently produced by all the noncoal mining operations in the United States every year. Indeed, Menzie acknowledged during his congressional testimony that these "increased environmental costs" and the "social scrutiny of and re-strictions on mineral resource extraction" might well limit global ability to meet the accelerating demand for copper.[6]

Even if all the necessary technology, energy, and political will were in place to extract the planet's copper in some reasonably acceptable manner, there may still be even harder limits to continued growth in resource consumption. Analyzing and predicting the possibilities of future mineral resource exhaustion is a notoriously tricky issue. As Daniel Jackling proved in 1906, and others have repeatedly proven since, new technologies can transform worthless rock into valuable ore and help geologists discover new deposits. Likewise, other metals or materials may be substituted when a mineral becomes scarce, such as aluminum for copper and steel, though there is often a cost in diminished performance or increased price. Given these variables, dire predictions of impending metal shortages in the past have sometimes proven overly pessimistic. In 1972, an international group of scientists and industrialists called the Club of Rome published *The Limits to Growth*, which forecast the collapse of industrial civilization in the next century due to the exhaustion of natural resources and environmental devastation.[7] Three decades later, debates continue about the book's methods and accuracy. Many experts, like the USGS geologist Menzie, argue the book's theoretical model was fundamentally flawed because it underestimated the ability of exploration and new technologies to add to the world's stock of mineral reserves.[8]

Perhaps more famously, in 1980 the economist Julian Simon entered into a wager with the entomologist Paul Ehrlich and several other environmental scientists over whether the prices of five key metals, including copper, would rise or fall over the next decade. Twelve years earlier, Ehrlich had predicted that global population growth would cause mass starvation by the 1970s and 1980s in a best-selling book, *The Population Bomb*.[9] Ehrlich's worst fears had yet to materialize in 1980, but he and several of his colleagues continued to believe that the greedy human appetite for natural resources would soon begin to overwhelm global supplies. Simon, on the other hand, bet that human inventiveness and discoveries would drive mineral prices down by 1990.[10]

Simon won. Ehrlich and his colleagues sent him a check for $576.07, which was the amount (adjusted for inflation) that the price of the five metals had gone down. Copper prices, for example, had dropped nearly 25 percent, while world production was up more than 27 percent.[11] The price of several of the other metals had dropped even more sharply. At the time, some viewed Simon's victory as an amusing demonstration of a serious point: neo-Malthusians like Ehrlich were wrong. Human creativity and drive would always find ways to overcome any natural limits to growth.

However, as Thomas Homer-Dixon, a Canadian expert on global resource issues, notes in a recent reassessment of the famous wager, "Mr. Ehrlich and his colleagues may have the last (grim) laugh." With the double threat of decreasing oil supplies and increasing global warming, the twentieth-century pattern of using cheap energy to extract more natural resources may no longer be affordable or sustainable. To be sure, humans will still find new ways to overcome some of these limits, Homer-Dixon argues, but "our species' capacity to innovate, and to deliver the fruits of that innovation when and where they're needed, isn't infinite."[12]

Other researchers have taken an even more fundamental look at the issue by attempting to estimate the globe's total reserves of potentially available copper and other metals. In a widely publicized 2006 article, a collaborative team of resource geologists and economists calculated that the amount of copper currently in use combined with all the known potentially exploitable deposits still in the earth constituted a total global copper supply of 1.6 billion metric tons. The team further estimated that to sustain the current average standard of living in North America requires the use of about 170 kilograms (375 pounds) of copper per person. Thus, if the entire projected population of the globe in 2050 (8.7 billion people) were to use as much copper as the average North American, it would require 1.7 billion metric tons of copper—about 100 million tons more than the total estimated global supply. To come even close to meeting the demand would require mining every known copper deposit on the planet—regardless of environmental, energy, or social costs—and diligently recycling all the copper currently in use so that not a single scrap was lost. To match the extraordinary American use of copper at just under 250 kilograms (550 pounds) per capita, however, would be all but impossible short of finding another planet to mine. In sum, the authors conclude, "virgin stocks of several metals appear inadequate to sustain the modern 'developed world' quality of life for all of Earth's people under contemporary technology."[13]

Still, there is always substitution, improved recycling, and the possibility that clever humans will still triumph in some unforeseeable way. As one Dutch chemist rightly noted, "You could go to the moon to mine precious minerals. The question is: could you afford it?"[14] Ultimately, there is simply no way to predict future supplies of copper or any other metal with certainty.[15] However, we can be quite certain of one thing. If we realistically hope to bring much of the world up to the standard of living enjoyed in the developed nations without devastating the global ecosystem, it will require deep and sustained investments in improved mining, recycling, and pollution control technologies, not to mention the clean- and sustainable-energy sources needed to fuel them. Whether the developed nations like the United States—with their metallic infrastructures largely in place and comparatively sustainable—will prove adequate to the challenge is a troubling question.

For the first time in a century, on March 28, 2008, the Clark Fork and Blackfoot rivers flowed freely through southwestern Montana. At noon that day, engineers breached the Milltown Dam just east of the city of Missoula, and the rivers (which merge just above the dam) began the gradual process of returning to their natural course. Joining a crowd of cheering onlookers were Montana's two U.S. senators and the state's governor. According to a local newspaper account, the government officials all agreed that the removal of the old hydroelectric dam was an important step forward that "represented Montana's shift from an extraction to a restoration economy, creating jobs that protect the environment and use the state's natural resources in a sustainable way rather than plundering them."[16]

Once a symbol of Montana's economic and technological prowess, the Milltown Dam had in more recent decades come to signify all that had been wrong with the state's more than century-long affair with copper mining. William Clark, one of the original Butte copper kings whose properties later became part of the Anaconda, had put the timber and stone dam into operation almost exactly a century earlier. The impressive brick powerhouse held three turbines that produced about 3,400 kilowatts, enough to electrify part of Missoula as well as supply the Anaconda's nearby Bonner mill, where workers cut timbers destined for Butte's underground.[17] To many, it all seemed to fit perfectly. Copper mining not only brought good jobs and economic growth. It also provided electricity and the red metal to carry it, helping to usher a far western outpost of the American continental empire into a dazzling new modern age. To be sure, there were the farmers and ranchers in the Deer Lodge Valley who complained that the smelter smoke killed their crops and stock animals, and there were the legions of men in Butte who died young in the mines or lived to be crippled later by silicosis. But for much of the twentieth century, many Montanans likely believed the benefits seemed to outweigh the costs. After all, Nick Bielenberg's dead cattle and Mount Moriah Cemetery's dead miners were 120 miles away, far removed from the modern city of Missoula with its pleasant streets and homes bathed in bright electric lights.

Then, in May 1981, a routine check by the Missoula City–County Health Department discovered that well water in the neighboring community of Milltown contained levels of arsenic five to ten times higher than deemed safe under federal drinking water standards. Clark's dam had been acting as a trap for arsenic and other heavy metals that, much to everyone's surprise, had washed down the river from Butte and Anaconda. Further tests revealed that the sediment behind Milltown Dam contained 1,800 tons of arsenic and tens of thousands of tons of other heavy metals. At that point, the groundwater contamination was still limited to the Milltown area. But like the river, the groundwater beneath flows toward Missoula, raising the possibility of contaminating the drinking water of more than fifty thousand people. Two years later, the Environmental Protection Agency designated the Milltown reservoir as the northern tip of the nation's largest Superfund site.[18] After another decade of research and debate, ARCO, the EPA, and the state agreed to an ambitious and costly plan to remove both the dam and about 2.2 million cubic yards of the heavy-metal-laced sediment. ARCO (or to be more specific, BP-ARCO following its merger with British Petroleum in 2000) is shouldering most of the $120 million cost of the project, one part of the Superfund liability it inherited after its 1977 purchase of Anaconda.

In October 2007, following the diversion of the river into a temporary channel, big front-end loaders began dumping the jet black Milltown sediment into a long line of twenty-eight brand-new railcars. Before the sediment-removal work is completed sometime in 2009, engineers estimate the loaders will fill about thirty thousand of the hundred-ton-capacity cars. From Milltown the train with its loaded cars retraces in reverse the route the tailings had followed down the 120-mile course of the Clark Fork, taking them back to their original sources at the headwaters near Butte and Anaconda. At a loading ramp on the edge of the 3,500-acre Opportunity tailings ponds outside Anaconda, an excavator transfers the sediment to giant forty-ton trucks, modern descendants of those once used to mine the Berkeley Pit. Big engines roaring, the trucks slowly inch up the gravel roads cut into the side of the forty-foot-tall tailings piles and dump their loads of sediment on top. Operators driving a pair of yellow bulldozers then spread the sediment out over about twenty acres, taking care to keep the layer about two feet thick. When the relocation project is completed, the toxic Milltown mud will cover about one third of the old

Opportunity tailings ponds, now renamed the BP-ARCO Waste Repository. Although the sediment contains many of the same heavy metals as
the tailings ponds, their concentration is far lower, and the Milltown mud
actually makes a reasonably good topsoil for capping the tailings. Almost
nothing will grow in the nearly sterile old tailings, but the Milltown sediment is rich in a century's worth of accumulated organic material. ARCO
and the EPA believe that the capping process will make the current dusty
desert there bloom, and that both the tailings and the Milltown sediment
will be far safer than before.[19]

Not everyone has been happy with the EPA's plan. Some of the residents
of the tiny town of Opportunity just over a mile from the newly christened
"waste repository" view the project as "adding insult to injury." "It's lovely
to have Missoula cleaned up," one former smelter worker said, "but we
don't want it here, either."[20] In the enthusiasm sparked by the breaching of
the dam in March 2008, though, such concerns seemed distant. The excitement only grew when just a few days after the breach scientists reported that a foot-long rainbow trout swam up the Clark Fork and past the
former dam site for the first time in more than a century. Fisheries biolo-
gists said the trout was the vanguard of tens of thousands of others to
come that spring, some of which will make their way up to spawn on the
Blackfoot, a river made famous by Norman Maclean's story of family and
fly-fishing, *A River Runs Through It*. Many journalists covering the dam
breach could not resist referencing the vaguely spiritual sentence in the
book from which Maclean took his title: "Eventually, all things merge into
one, and a river runs through it."[21]

For some Montanans, the dam removal and the trout symbolized the
promise of a new Montana and a New West. Pat Williams, a former U.S.
congressman from the state who now holds the intriguing title of Northern
Rockies Director of Western Progress, believes the trout represented "the
flipside of the industrial revolution: new opportunities for clean development and recreation, salaries and benefits for our workers, and profits for
our businesses." Williams predicts the coming of a new "Restoration Economy" where "yesterday's scars on our landscape are today's pay dirt" and the
rewards include "an improved environment with trails, parks, rivers running free, and trout able once again to follow their ancient instincts."[22]

So it seemed that a New West could emerge from the smoke and tailings
of the old, one where free-swimming rainbow trout and economic growth
might peacefully coexist, and the human and environmental sacrifices of
earlier generations would no longer be necessary. Indeed, the Milltown

project provides a fitting reminder that machines descended from those Daniel Jackling and the Anaconda used to make their open-pit copper mines could be just as efficient at cleaning up the environment as destroying it. Likewise, engineers and scientists in Butte are now in the forefront of efforts to find new technologies capable of profitably cleaning polluted mine water by extracting the valuable metals—in a sense, much improved versions of the concentration and flotation systems Jackling perfected to extract the copper from the low-grade Bingham ore. Recently, industrial ecologists have estimated that there are about eighty-five million metric tons of copper in landfills (hidden in refrigerators, radios, blenders, and other abandoned consumer goods) across North America as well as in mine dumps like the BP-ARCO Waste Repository near Opportunity. Were some future Daniel Jackling to perfect a means of profitably processing entire landfills in order to separate out the minuscule amounts of copper and other valuable metals, it might one day prove possible to develop these "technospheric mines."[23]

Given the history chronicled in this book, however, any predictions of future breakthroughs and painless techno-fixes must be viewed with considerable skepticism. The modernist promise of infinite progress through science and technology has repeatedly underestimated the complexity of technological and environmental systems, thus creating a host of new problems with every supposed solution. Absent just such technological breakthroughs, however, much of the new copper needed to bring electric light and power to the village of Chakai Haat and others like it around the globe will almost certainly have to come from big low-grade open-pit copper mines using Jackling's well-proven methods of mass destruction mining.[24] There are, to be sure, different approaches that could be followed. Americans and others living in industrialized nations could drastically diminish their use of copper and other minerals in order to share the existing supplies with the world. Tighter controls on population growth could significantly decrease the number of potential mineral consumers in 2050 and beyond. Modernizing countries could reject the extremes of Western consumerist culture in favor of less resource-intensive economies and lives. At this writing, none of these alternatives seems very likely. As in the past, humans will be sorely tempted to maintain and expand their prosperity by using mass destruction technologies that shift many of the costs to the environment. Ironically, then, the best hope for minimizing environmental degradation may be to actually increase our commitment to the very scientific and technological tools that have caused so many problems over the past century. Perhaps a more humble, nuanced, and complex approach to understanding and managing the intricate envirotechnical system that we all live in today will afford better results than those achieved by our predecessors.

Given all of these realities, though, there was something troubling about the almost vindictive glee with which Montanans celebrated the destruction of Milltown Dam and the prospect of creating Pat Williams's post-industrial "Restoration Economy." To be sure, the toxic mine wastes at the site had to be cleaned up, and the start of the remediation project after many battles and delays was a moment well worth celebrating. In all of the understandable excitement over reunited rivers and free-swimming rainbows, however, precious few seemed to recognize how the dirty upstream copper mines had once helped to wire Montana and America, or how the dam had generated relatively clean electric power that streamed over those copper lines into the homes and businesses of Missoula.[25] Instead, many Montanans seemed to suggest that their future would henceforth be unmoored from such pedestrian industrial and technological realities. Theirs would be a pristine place apart, an environmental and recreational paradise where the roar of shovels and the whine of chainsaws would be heard only where profits might be had cleaning up the messes made by a dark and now best forgotten industrial past.

This trend is not unique to Montana. In a 1998 article in the *New York Times*, the author identifies "an emerging truism of the new West: the only good miner is a dead miner, preferably a quaint one of the nineteenth century." In much of the New West, residents of former mining districts welcome tourists to visit their historic Old West mines while staunchly opposing all new mineral development. In the new western economy, they argue, the value of the region is in terrestrial scenery, not subterrestrial resources. As one Montana economist suggests, "The open pit mine or clearcut mountain top that once provided economic security for a community now threatens the influx of future jobs."[26]

Some critics have derided the growth of such a service-based tourist economy; one writer describes it as the "clownish and implicit servility that seems increasingly to color the vacationland of western Montana."[27] With recent sharp increases in commodity prices, a few mining operations have also nonetheless begun to quietly revive around the West.[28] Likewise, the pits at Bingham, Morenci, Chino, and elsewhere continue to provide their copper to the nation. Still, the more difficult question about global mineral supplies has generally remained unasked: where will the new supply of metals needed to bring the majority of the world up to Western living standards come from? As was the case a century ago, most of the accessible mineral deposits in the United States are still in the western mountains, though the more easily developed deposits have already been mined. In stark contrast to a century ago, however, today the United States feeds its still growing appetite for metals by importing more minerals than it exports. In 2006, the nation imported over a million tons of refined copper—ten times the amount it exported to the rest of the world. Contrast this to 1906, the year Jackling opened his Bingham Pit, when the nation exported more than fifteen times as much copper as it imported. Indeed, since 1976, the United States has always imported far more of the world's supply of copper than it has given back.[29]

As one resource geologist has recently argued, in an ideal egalitarian world all countries would "share the burden of supplying the world's population with minerals rather than expect others to shoulder the environmental burden."[30] For much of the twentieth century, the mass destruction technologies perfected by Jackling and others ensured that Americans would reap the benefits as well as the costs of cheap and abundant copper—both a wired nation and a domestic landscape scarred with dead zones. During the last three decades, though, the United States has increasingly displaced the consequences of its voracious mineral appetites onto other nations. Some in places like Bingham, Butte, Anaconda, and many other areas of the American West might reasonably argue that their lands and communities have already paid more than their fair share of the costs of national development. However, so long as westerners and other Americans continue to increase their consumption of the world's finite supplies of copper and other metals, such arguments must always ring slightly hollow.

Unlike the Milltown Dam, Daniel Jackling's Woodside mansion outside of San Francisco still stands for the moment. At this writing, the preservationists appear to have succeeded in stopping Steven Jobs from tearing the mansion down, and he continues to wait for someone willing to pay the millions needed to disassemble the house and rebuild it on another site. Meanwhile, the house remains locked in its moldering anonymity, the history of the man who built it only briefly to be recalled whenever the next chapter in Jobs's real estate saga excites the public's fleeting interest. Such is the New West in the early twenty-first century. The region, and indeed the nation as a whole, has yet to escape the destructive beliefs of the past that divided the human from the natural, the technological from the ecological, and the mine from the consumer. Instead, the dividing lines seem to have merely been redrawn elsewhere, between a handful of sheltered enclaves of affluence and the remaining majority of the world from which the requisite raw materials to feed that affluence would come. On one side, humans forgot the material sources of their civilization and rushed to tear down the dams, houses, and other symbols of what they viewed as an irrelevant and simplistically evil past. On the other, humans increasingly lived in a world that looked very much like the one their wealthy neighbors were working so diligently to erase. In time, these new dichotomous divisions will surely prove as illusory and dangerous as those of the past.

In 1944, a historian could still boast without embarrassment that Jackling and the other American mining geniuses of the age "not only shook the earth in passing, they shifted it, tossed it about literally, creating molehills out of mountain ranges."[31] For all the reasons chronicled in this book, few today would strike such an unapologetic tone of celebration. But we should also take care not to indulge in a new type of hubristic and all too modern boast, one in which we prematurely declare American independence from the costs of industrialization and natural resource extraction while the vast majority of the world remains desperate to share in even a small part of the extraordinary mineral wealth enjoyed in this nation. So long as the shantytowns of New Delhi and villages around the world fall dark at night, any future New West should not shrink from its moral obligation to share its still considerable reserves of mineral wealth with the world while sharply reducing its own consumption of current supplies. Rather than fleeing their mineral past in pursuit of a dangerous and illusory dream of pristine landscapes divorced from the technological foundations of modern existence, citizens of the New West would do well to reconsider the economic, global, and moral benefits of leading an effort to develop viable alternatives to the technologies and culture of mass destruction. The first step might prove the most difficult: rejecting the pernicious divisions of modernity and instead learning to see humans and their technologies as entirely natural and inextricable parts of nature. But if the New West and the nation as a whole can make that leap, they might yet lead the way in creating a culture and society built upon an environmentally sustainable system of mining, manufacturing, consumption, and recycling.

That would be a challenge well worth the efforts of a new generation of Jacklings and Cottrells determined to provide the world with the copper it needs, as well as abundant clean rivers teeming with free-swimming rainbow trout.

Notes

ONE. IN THE LANDS OF MASS DESTRUCTION

1. Andrea Gemmet, "The House That Jackling Built," *Almanac*, 27 February 2002.

2. Andrea Gemmet, "Steve Jobs Wins Fight to Tear Down Woodside House," *Almanac*, 22 December 2004. The headline refers to the earlier town council decision later overturned by the Superior Court judge. See also Patricia Leigh Brown, "In Silicon Valley, Tear-Down Interrupted," *New York Times*, 15 July 2004.

3. Patty Fisher, "Historic? By Whose Definition?" *Mercury News*, 18 January 2006.

4. Letter from Anthea M. Hartig, Chair, California State Historical Resources Commission to Paul Goeld, Mayor, Woodside, California, 15 September 2004, available online at www.friendsofthejacklinghouse.org/letter1.html, accessed 9 September 2007.

5. Ibid.

6. See, for example, Ira Beaman Joralemon, *Romantic Copper: Its Lure and Lore* (New York and London: D. Appleton-Century, 1934); Watson Davis, *The Story of Copper* (New York and London: Century, 1924), A. B. Parsons and American Institute of Mining Metallurgical and Petroleum Engineers, *The Porphyry Coppers* (New York: AIMMPE, 1933), and T. A. Rickard, *Man and Metals: A History of Mining in Relation to the Development of Civilization* (New York and London: Whittlesey House, McGraw-Hill, 1932).

7. Quoted in Gemmet, "The House That Jackling Built."

8. Ronald C. Brown, "Daniel C. Jackling and Kennecott: A Mining Entrepreneur's Adjustment to Corporate Bureaucracy," *Mining History Journal* 10 (2003): 133.

9. Douglas Brinkley, *Wheels for the World: Henry Ford, His Company, and a Century of Progress* (New York: Penguin Books, 2003), 517.

10. "Henry Ford Is Dead at 83 in Dearborn," *New York Times*, 8 April 1947.

11. *Utah History Encyclopedia*, s.v. "Daniel Cowan Jackling."

12. In this emphasis on speed and nonselectivity, the concept of mass destruction is very different than the idea of "brute force technology" offered in Paul R. Josephson, *Industrialized Nature: Brute Force Technology and the Transformation of the Natural World* (Washington, D.C.: Island Press/Shearwater Books, 2002).

13. William Cronon, *Nature's Metropolis: Chicago and the Great West* (New York: Norton, 1991), 256–257.

14. This point is well made in Bruno Latour, *We Have Never Been Modern* (Cambridge: Harvard University Press, 1993), 10–11.

15. William Cronon, "The Trouble with Wilderness; or, Getting Back to the Wrong Nature," *Environmental History* 1 (1996): 7–28.

16. "Ancient Lead Emissions Polluted Arctic," *Science News*, 5 November 1994.

17. Duncan Adams, "Did Toxic Stew Cook the Goose?" *High Country News*, 11 December 1995; and Mark Levine, "As the Snake Did Away with the Geese," *Outside* (September 1996).

18. Edwin Dobb, "Pennies from Hell: In Montana, the Bill for America's Copper Comes Due," *Harper's* (October 1996): 42–53.

19. Levine, "As the Snake Did Away with the Geese."

20. Jeffrey St. Clair, "Something About Butte," *Counterpunch* (January 2003).

21. This powerful Montana and later international mining company has gone through several corporate forms and names, including a period as a part of the Standard Oil trust when it was the Amalgamated Copper Company. For the sake of convenience and brevity, in this book the corporation will be referred to simply as the Anaconda.

22. Mumford's fascination with the Deutsches Museum mines is related in Rosalind Williams, *Notes on the Underground: An Essay on Technology, Society, and the Imagination* (Cambridge: MIT Press, 1990), 4–5.

23. Lewis Mumford, *Technics and Civilization* (1934; New York: Harcourt Brace, 1963), 65–77, quote from 69–70.

24. Williams, *Notes on the Underground*, 7–8.

25. Richard White, "From Wilderness to Hybrid Landscapes: The Cultural Turn in Environmental History (American West Portrayals)," *Historian* 66, no. 3 (2004): 560.

26. See, for example, T. C. Onstott et al., "The Deep Gold Mines of South Africa: Windows into the Subsurface Biosphere," *Proceedings of the SPIE* 3111 (1997): 344–357.

27. Edmund Russell, for example, has suggested the provocative thesis that stock animals, dogs, and perhaps other organisms can be usefully analyzed as human-designed technologies, which inspired a conference on this theme held at the Hagley Museum and Library. See Susan R. Schrepfer and Philip Scranton, eds., *Industrializing Organisms: Introducing Evolutionary History* (New York: Routledge, 2004). See also Michael Bess, "Artificialization and Its Discontents," and

Angela Gugliotta, "Environmental History and the Category of the Natural," both in *Environmental History* 10 (January 2005). Also seminal have been Arthur F. McEvoy, "Working Environments: An Ecological Approach to Industrial Health and Safety," *Technology and Culture* 36 (Supplement, 1995): S145–172; Christopher Sellers, "Body, Place and the State: The Makings of An 'Environmentalist' Imaginary in the Post–World War II U.S.," *Radical History Review* 74 (1999): 31–64; Conevery Bolton Valencius, *The Health of the Country: How American Settlers Understood Themselves and Their Land* (New York: Basic Books, 2002); Gregg Mitman, "In Search of Health: Landscape and Disease in American Environmental History," *Environmental History* 10 (2005): 184–210; and Linda Nash, "Finishing Nature: Harmonizing Bodies and Environments in Late-Nineteenth-Century California," *Environmental History* 8 (2003): 25–52.

From a history of science perspective, Bruno Latour's concept of a web of interconnected actors in which nature, humans, machines, maps, etc., are all linked avoids simple dichotomies of the natural and the artificial. See Bruno Latour, *Science in Action: How to Follow Scientists and Engineers Through Society* (New Haven: Yale University Press, 1988), and idem, *We Have Never Been Modern* (Cambridge: Harvard University Press, 1993). Examples of related (though far from identical) arguments emerged among environmental historians like William Cronon with his now classic article, "The Trouble with Wilderness; or, Getting Back to the Wrong Nature," *Environmental History* 1 (1996): 7–28; Richard White, *The Organic Machine: The Remaking of the Columbia River* (New York: Hill and Wang, 1996); and Mark Fiege, *Irrigated Eden* (Seattle: University of Washington Press, 1999). A pathbreaking early work was Donna Haraway's "A Cyborg Manifesto: Science, Technology, and Socialist-Feminism in the Late Twentieth Century," reprinted in *Simians, Cyborgs, and Women: The Reinvention of Nature* (New York; Routledge, 1991), 149–181.

28. See Tim LeCain, "The SHOT/HSS Roundtable on Envirotech Themes," *Envirotech Newsletter* 5, no. 2 (Fall 2005), 1–3, available online at http://www .stanford.edu/~jhowe/Envirotech/newsletters.html#Archived%20Newsletters, accessed 7 June 2007. Others on the panel—which in addition to Russell included John Staudenmaier, Martin Reuss, and moderator/organizer Hugh Gorman—made similar points about the convergence of technological and environmental systems.

29. Cronon, *Nature's Metropolis*, xix.

TWO. BETWEEN THE HEAVENS AND THE EARTH

1. Donald MacMillan, "A History of the Struggle to Abate Air Pollution from Copper Smelters of the Far West, 1885–1933" (Ph.D. diss., University of Montana, 1973), 111.

2. Robert Friedel, "New Light on Edison's Light," *American Heritage of Invention and Technology* (Summer 1985): 22–27.

3. Thomas P. Hughes, "Edison's Method," in *Technology at the Turning Point*, ed. W. B. Pickett (San Francisco: San Francisco Press, 1977), 9.

4. Association of Edison Illuminating Companies, Committee on St. Louis Exposition, *"Edisonia"* (New York, 1904), 49–50.

5. Ibid.

6. Frank Lewis Dyer and Thomas Commerford Martin, *Edison, His Life and Inventions* (New York and London: Harper & Brothers, 1910), 626.

7. On American electrification, see Thomas Parke Hughes, *Networks of Power: Electrification in Western Society, 1880–1930* (Baltimore: Johns Hopkins University Press, 1983), 18–46; David E. Nye, *Electrifying America : Social Meanings of a New Technology, 1880–1940* (Cambridge: MIT Press, 1990); idem, "Electrifying the West, 1880–1940," *European Contributions to American Studies* 16 (1989): 183–202; Warren D. Devine Jr., "From Shafts to Wires: Historical Perspectives on Electrification," *Journal of Economic History* 43 (1983): 347–372; Richard F. Hirsh, *Technology and Transformation in the American Electric Utility Industry* (Cambridge: Cambridge University Press, 1989); and Harold I. Sharlin, *The Making of the Electrical Age: From the Telegraph to Automation* (London: Abelard-Schuman, 1963).

8. Harold I. Sharlin, "Electrical Generation and Transmission," in *Technology in Western Civilization*, ed. Melvin Kranzberg and Carroll W. Pursell (New York: Oxford University Press, 1967), 579.

9. Watson Davis, *The Story of Copper* (New York: Century, 1924), 257.

10. Bureau of Mines, *1952 Materials Survey: Copper* (Washington, D.C.: National Security Resources Board, 1952), S-1.

11. D. C. Jackling, "Copper—The Everlasting Metal," *Mines Magazine* 27 (November 1937): 15.

12. Davis, *Story of Copper*, xii.

13. Lewis Mumford, *Technics and Civilization* (1934; New York: Harcourt Brace, 1963), 70, 221–224, 229–230, 255.

14. David E. Nye, *American Technological Sublime* (Cambridge: MIT Press, 1994).

15. Charles Henry Janin, "The Copper Situation," a paper presented to the Commonwealth Club of California, 26 June 1924, Charles H. Janin Collection, folder 13, box 15, Huntington Library, San Marino, California.

16. Thomas A. Rickard, *Interviews with Mining Engineers* (San Francisco: Mining and Scientific Press, 1922), 215; "Bear Lake Reservoir and Lifton Pumping Plant of the Utah Power and Light Company," D. C. Jackling Collection, box 3, Bancroft Library; John S. McCormick, "The Beginning of Modern Electric Power Service in Utah, 1912–22," *Utah Historical Quarterly* 56 (1988): 4–22.

17. Carrie Johnson, "Electrical Power, Copper, and John D. Ryan," *Montana: The Magazine of Western History* 38 (1988): 24–37.

18. "Overland by Electric Cars Near Reality: Men Behind Biggest Power Supplying Corporations in West Also Control Mines That Supply Copper Required for Electrification," *Salt Lake City Tribune*, 19 January 1912.

19. A. B. Parsons, *The Porphyry Coppers in 1956* (New York: American Institute of Mining, Metallurgical, and Petroleum Engineers, 1957), 27.

20. Davis, *Story of Copper*, 246; Christopher J. Schmitz, "The Changing Structure of the World Copper Market, 1870–1939," *Journal of European Economic History* 26 (1997): 296–297 and n. 4.

21. Kenneth Barbalace, "Periodic Table of the Elements," available online at http://EnvironmentalChemistry.com/yogi/periodic/, accessed 10 October 2007.

22. Donatella Romano and Francesca Matteucci, "Contrasting Copper Evolution in ω Centauri and the Milky Way," *Monthly Notices of the Royal Astronomical Society*, Letters 378 (1): L59–L63. For a less technical explanation of Romano and Francesca's work, see Ken Croswell, "The Stellar Origin of Copper," available online at http://kencroswell.com/Copper.html, accessed 15 February 15 2008.

23. Croswell, "Stellar Origin of Copper."

24. Ira Joralemon, *Copper: The Encompassing Story of Mankind's First Metal* (Berkeley: Howell-North Books, 1973), 202; Christopher H. Gammons, John J. Metesh, and Terrence E. Duaime, "An Overview of the Mining History and Geology of Butte, Montana," *Mine Water and the Environment* 25 (2007): 70–72.

25. The extraordinary 10,000-mile figure is confirmed by the Montana Bureau of Mines and Geology, but it does include *all* underground passageways, including the many small vertical "man ways" that allowed miners to move between levels. See Barbara LaBoe, "New Map Plots Butte Underground," *Montana Standard*, June 7 2004.

26. Walter H. Weed, "Geology and Ore Deposits of the Butte District, Montana," *U.S. Geological Survey Professional Paper 74* (Washington, D.C.: Government Printing Office, 1912), 50.

27. The film is *Evel Knievel* (1971) and features George Hamilton in the starring role. On ground subsidence, see "Superfund Redevelopment Pilots, Silver Bow Creek/Butte Area," U.S. Environmental Protection Agency Fact Sheet, July 2000.

28. James C. Scott, *Seeing Like a State: How Certain Schemes to Improve the Human Condition Have Failed* (New Haven: Yale University Press, 1998), 183.

29. Quoted in Kent Curtis, "An Ecology of Industry: Mining and Nature in Western Montana, 1860–1907" (Ph.D. diss., University of Kansas, 2001), 95. See also Eric Nystrom, "Mapping Underground Drifton: The Evolution of Underground Mine Maps," *Canal History and Technology Proceedings* 25 (2006): 79–96.

30. Andrew C. Isenberg, *Mining California: An Ecological History* (New York: Hill and Wang, 2005), 34–35.

31. Isenberg, *Mining California*, 24–25.

32. Ibid., 51.

33. Otis E. Young Jr., *Western Mining: An Informal Account of Precious-Metals Prospecting, Placering, Lode Mining, and Milling on the American Frontier from Spanish Times to 1893* (Norman: University of Oklahoma Press, 1970), 247.

34. Rodman W. Paul, *Mining Frontiers of the Far West, 1848–1880* (New York: Holt, 1963), 64. See also, Young, *Western Mining*, 247.

35. Paul, *Mining Frontiers*, 68.

36. Ibid., 70–71; Martin Lynch, *Mining in World History*, (London: Reaktion, 2003), 152.

37. Michael P. Malone and Richard B. Roeder, *Montana: A History of Two Centuries* (Seattle: University of Washington Press, 1976), 202.

38. Paul, *Mining Frontiers*, 147; Malone, *Montana*, 202–204; Isaac Frederick Marcosson, *Anaconda* (New York: Dodd, 1957), 41–42.

39. Marcosson's dramatic account in *Anaconda*, 50–51, emphasizes Tevis's doubts about the mine.

40. Paul, *Mining Frontiers*, 146–147; Malone, *Montana*, 204–205.

41. Marcosson, *Anaconda*, 40–41.

42. Ibid., 61; *Bulletin of the American Institute of Mining Engineers*, May 1915, xxviii.

43. Michael Punke, *Fire and Brimstone: The North Butte Mining Disaster of 1917* (New York: Hyperion, 2006), 22.

44. Bruce Braun, "Producing Vertical Territory: Geology and Government in Late Victorian Canada," *Ecumene* 7 (2000): 22–23.

45. F. A. Linforth, "Application of Geology to Mining in the Ore Deposits at Butte Montana," in *Ore Deposits of the Western States* (New York: American Institute of Mining and Metallurgical Engineers, 1933), 700.

46. "The Knowles Steam-Pumps," *Manufacturer and Builder* 8 (January 1876): 4–5.

47. Young, *Western Mining*, 150; Curtis, "Ecology of Industry," 115–118.

48. Alan Derickson, *Worker's Health, Worker's Democracy: The Western Miners' Struggle for Health and Safety* (Ithaca: Cornell University Press, 1988).

49. *Iron and Coal Trade Review*, 29 October 1920, 587.

50. Kenneth Ackerman, "Deep-Sea Diving a Century Ago," *American Heritage of Invention & Technology* 4 (1994): 58–63.

51. G. W. Grove, *Self-Contained Oxygen Breathing Apparatus: A Handbook for Miners* (Washington, D.C.: Bureau of Mines, 1941), 6–11; E. H. Denny and M. W. von Bernewitz, *Value of Oxygen Breathing Apparatus to the Mining Industry* (Washington, D.C.: Bureau of Mines, 1923).

52. James W. Paul, *The Use and Care of Mine-Rescue Breathing Apparatus* (Washington, D.C.: Bureau of Mines, 1911), 1–5.

53. Ibid., 10–13; Punke, *Fire and Brimstone*, 47–59.

54. Grove, *Oxygen Breathing Apparatus*, 6–11.

55. Denny, *Value of Oxygen Breathing Apparatus*, 4.

56. Thomas Thornton Read, *The Development of Mineral Industry Education in the United States* (New York: American Institute of Mining Engineers, 1941), 258; F. W. Sperr, "The National Mine Safety Demonstration," *Bulletin of the Society for the Promotion of Engineering Education* 2 (1912): 231–241.

57. C. A. Mitke, "Method of Testing Draeger Oxygen Helmets at the Copper Queen Mine," *Transactions of the American Institute of Mining Engineers* 47 (1914):

78–81. On the Phelps Dodge fire, see Joseph P. Hodgson, "Mining Methods at the Copper Queen Mines," *Transactions of the American Institute of Mining Engineers* 44 (1914): 324.

58. Brian Shovers, "The Perils of Working in the Butte Underground: Industrial Fatalities in the Copper Mines, 1880–1920," *Montana: The Magazine of Western History* (Spring 1987): 26–39.

59. Recent work placing the human body at the center of historical environmental analysis is especially intriguing in this light. See the collection of articles in Greg Mitman, Michelle Murphy, and Christopher Sellers, eds., *Landscapes of Exposure: Knowledge and Illness in Modern Environments* (Chicago: University of Chicago Press, 2004).

60. Malone, *Montana*, 271–275. See also Michael P. Malone, *The Battle for Butte: Mining and Politics on the Northern Frontier, 1864–1906* (Seattle: University of Washington Press, 1981).

61. This process of technological and managerial deskilling of workers in other industries is well explained in Harry Braverman, *Labor and Monopoly Capital: The Degradation of Work in the Twentieth Century* (New York: Monthly Review Press, 1975). See also David F. Noble, *Forces of Production: A Social History of Industrial Automation* (New York: Knopf, 1984).

62. Punke, *Fire and Brimstone*, 8–15.

63. Malone, *Montana*, 274. See also Punke, *Fire and Brimstone*.

64. Malone, *Montana*, 274–275.

65. Punke, *Fire and Brimstone*, 2.

66. Ibid.

67. Malone, *Montana*, 274–275.

68. Charles Perrow, *Normal Accidents: Living with High-Risk Technologies* (Princeton: Princeton University Press, 1999).

69. Punke, *Fire and Brimstone*, 10–11.

70. Scott, *Seeing Like a State*. Of course, Scott's work is focused on governmental schemes, not corporate ones. Yet his basic theoretical points about legibility, simplification, and governability can be usefully applied to some large-scale corporate-sponsored projects.

71. Rossiter W. Raymond to Sarah, 10 June 1869, acc. 1063, Rossiter Worthington Raymond Collection, Colorado Historical Society, Denver.

72. Raymond to Sarah Raymond, 24 August 1872, Rossiter Worthington Raymond Collection.

73. T. A. Rickard, *The Utah Copper Enterprise* (San Francisco: Abbott Press, 1919), 38.

74. Peter J. Bowler, *The Earth Encompassed: A History of the Environmental Sciences* (New York: Norton, 1992), 377.

75. *Prospector*, 1937, Colorado School of Mines, 51.

76. *Oredigger*, 4 September 1923.

77. *Oredigger*, 25 March 1924.

78. Gail Bederman, *Manliness and Civilization: A Cultural History of Gender and Race in the United States, 1880–1917* (Chicago: University of Chicago Press, 1996).

79. Michael L. Smith, *Pacific Visions: California Scientists and the Environment, 1850–1915* (New Haven: Yale University Press, 1987), 76–96.

80. Jesse Robert Morgan, *A World School: The Colorado School of Mines* (Denver: Sage Books, 1955), 109–111.

81. *Oredigger*, 29 March 1927.

82. "Women in Engineering," *Oredigger*, 14 December 1926.

83. Mary Midgley, *Science as Salvation: A Modern Myth and Its Meaning* (New York: Routledge, 1992).

84. David F. Noble, *A World Without Women: The Christian Clerical Culture of Western Science* (New York: Knopf, 1992); idem, *The Religion of Technology* (New York: Knopf, 1997).

85. Judy Wajcman, *Feminism Confronts Technology* (University Park: Pennsylvania State University Press, 1991), 143.

86. Sally Hacker, *Pleasure, Power and Technology* (Boston: Unwin Hyman, 1989); idem, "The Culture of Engineering: Woman, Workplace and Machine," *Women's Studies International Quarterly* 4 (1981): 341–353. See also Ruth Oldenziel, *Making Technology Masculine: Men, Women, and Modern Machines in America, 1870–1945* (Amsterdam: Amsterdam University Press, 2004); Arwen Palmer Mohun, *Steam Laundries: Gender, Technology, and Work in the United States and Great Britain, 1880–1940* (Baltimore: Johns Hopkins University Press, 2002); and Nina E. Lerman, Ruth Oldenziel, and Arwen Mohun, *Gender and Technology: A Reader* (Baltimore: Johns Hopkins University Press, 2003).

87. Mark David Spence, *Dispossessing the Wilderness: Indian Removal and the Making of the National Parks* (New York: Oxford University Press, 1999).

88. Karl Jacoby, *Crimes Against Nature: Squatters, Poachers, Thieves, and the Hidden History of American Conservation* (Berkeley: University of California Press, 2001).

89. Conevery Bolton Valencius, *The Health of the Country: How American Settlers Understood Themselves and Their Land* (New York: Basic Books, 2002).

90. Linda Nash, *Inescapable Ecologies: A History of Environment, Disease, and Knowledge* (Berkeley: University of California Press, 2006).

91. "Electrical Show Opened by Edison," *New York Times*, 12 October 1911.

92. See the letter to the editor, Nathan Shalit, *New York Times*, 6 July 1997.

93. Paul Israel, *Edison: A Life of Invention* (New York: John Wiley, 1998), 440–446.

THREE. THE STACK

1. Walter H. Weed, "Geology and Ore Deposits of the Butte District, Montana," *U.S. Geological Survey Professional Paper 74* (Washington, D.C.: Government Printing Office, 1912), 73–82.

2. See the explanation of what constitutes ore in H. Ries, *Economic Geology* (New York: John Wiley & Sons, 1947), 347.

3. Donald MacMillan, "A History of the Struggle to Abate Air Pollution from Copper Smelters of the Far West, 1885–1933" (Ph.D. diss., University of Montana, 1973), 22.

4. Ibid., 21.

5. Environmental Protection Agency, *Compilation of Air Pollution Emission Factors, Volume 1: Stationary Point and Area Sources* (Washington, D.C.: Government Printing Office, 1995), section 12.3-3.

6. James E. Fells, *Ores to Metals: The Rocky Mountain Smelting Industry* (Lincoln: University of Nebraska Press, 1979), 27–30, 273–274.

7. Donald M. Levy, *Modern Copper Smelting* (London: Charles Griffin, 1912); Edward Dyer Peters, *The Practice of Copper Smelting* (New York: McGraw-Hill, 1911).

8. Edmund Newell, "Atmospheric Pollution and the British Copper Industry, 1690–1920," *Technology and Culture* 38 (1997): 660; M.-L. Quinn, "Early Smelter Sites: A Neglected Chapter in the History and Geography of Acid Rain in the United States," *Atmospheric Environment* 23 (1989): 1281–1292.

9. Charles K. Hyde, *Copper for America* (Tucson: University of Arizona Press, 1998), 104. These figures are for total Montana production of copper, but the vast majority of this was from Butte.

10. Weed, "Geology and Ore Deposits," 20–21; Robert G. Raymer, *A History of Copper Mining in Montana* (Chicago: Lewis Publishing, 1930), 67.

11. MacMillan, "Struggle to Abate Air Pollution," 108–112, quote from 108.

12. Hyde, *Copper for America*, 94–100.

13. Ibid., 99–101.

14. The figures cited here, and others elsewhere in the book, are based on changes in the gross domestic product and the consumer price index, but the number could go as high as $32 million if based on changes in, for example, the unskilled wage level.

15. Edward Dyer Peters, *Modern Copper Smelting*, 10th ed. (New York: Scientific Publishing, 1895), 104–107.

16. Ibid., 106.

17. Donald MacMillan, *Smoke Wars: Anaconda Copper, Montana Air Pollution, and the Courts, 1890–1924* (Helena: Montana Historical Society Press, 2000), 90–92.

18. Hyde, *Copper for America*, 102.

19. Gorden Morris Bakken, "Was There Arsenic in the Air? Anaconda Versus the Farmers of the Deer Lodge Valley," *Montana* 41 (1991): 39–41.

20. Fredric L. Quivik, "Smoke and Tailings: An Environmental History of Copper Smelting Technologies in Montana, 1880–1930" (Ph.D. diss., University of Pennsylvania, 1998), 309.

21. The principal example of this interpretation can be found in MacMillan,

"Struggle to Abate Air Pollution." See also *Smoke Wars*, the posthumously published book based on this doctoral dissertation.

22. Quivik, "Smoke and Tailings," 320–321.

23. See, for example, Peter Dorman, "Environmental Protection, Employment, and Profit: The Politics of Public Interest in the Tacoma/Asarco Arsenic Dispute," *Review of Radical Political Economics* 16 (1984): 151–176.

24. Charles H. Fulton, *Metallurgical Smoke* (Washington, D.C.: Bureau of Mines, 1915), 83–84; Robert E. Swain, "Smoke and Fume Investigations," *Industrial and Engineering Chemistry* 41 (1949): 2385.

25. Peters, *Modern Copper Smelting*, 107.

26. M. L. Quinn, "Industry and Environment in the Appalachian Copper Basin, 1890–1930," *Technology and Culture* 34 (1993): 592.

27. Ibid., 575–612, quote from 611.

28. John E. Lamborn, "The Substance of the Land: Agriculture v. Industry in the Smelter Cases of 1904 and 1906," *Utah Historical Quarterly* 53 (1985): 310–312.

29. Ibid., 308–325; Fulton, *Metallurgical Smoke*, 82–83; MacMillan, "Struggle to Abate Air Pollution," 311–312; Bakken, "Was There Arsenic in the Air?" 34. On the technology of bag houses, see Fulton, *Metallurgical Smoke*, 49, and Peters, *Copper Smelting*, 578–597.

30. Joseph A. Blum, "South San Francisco: The Making of an Industrial City," *California History* 63 (1984): 128.

31. "Selby Plant Makes Liquid SO_2 from Waste Smelter Gases," *Engineering and Mining Journal* 150 (1949): 138–142.

32. Joseph A. Holmes, *First Annual Report of the Director of the Bureau of Mines* (Washington, D.C.: Bureau of Mines, 1910), 44.

33. Joseph A. Holmes et al., *Report of the Selby Smelter Commission*, Bulletin 98 (Washington, D.C.: Bureau of Mines, 1915).

34. Ibid., xvii.

35. Ibid., 39–51.

36. Ibid., 10.

37. Ibid., 363.

38. Approximately 10 to 20 percent of adults are hypersensitive to sulfur dioxide, while others build up a tolerance to the gas and show few apparent ill-effects from long-term exposure. See Otto Franzle, *Contaminants in Terrestrial Environments* (Berlin: Springer-Verlag, 1993), 224–227; David F. Tver, *Dictionary of Dangerous Pollutants, Ecology, and Environment* (New York: Industrial Press, 1981), 311; William H. Hallenbeck and Kathleen M. Cuningham-Burns, *Pesticides and Human Health* (New York: Springer-Verlag, 1985), 128.

39. James C. Scott, *Seeing Like a State: How Certain Schemes to Improve the Human Condition Have Failed* (New Haven: Yale University Press, 1998), 264.

40. Charles Lyons, *Fifteenth Annual Report of the Director of the Bureau of Mines* (Washington, D.C.: Bureau of Mines, 1925), 16.

41. Fulton, *Metallurgical Smoke*, 7.

42. Quivik, "Smoke and Tailings," 335.

43. Ibid., 310–312.

44. Frank Cameron, *Cottrell: Samaritan of Science* (New York: Doubleday, 1952), 38.

45. Cameron, *Cottrell*, 53–54; *Encyclopedia of American Biography*, s.v. "Frederick G. Cottrell."

46. John W. Servos, *Physical Chemistry from Ostwald to Pauling* (Princeton: Princeton University Press, 1990), 3–4, 53–70.

47. Robert E. Swain, "Smoke and Fume," 2384, credits Wislicenus with "being widely known for his work on injury to plant life by air-borne wastes." However, Wislicenus was primarily an organic chemist, and his smoke research was only a small part of his scientific work. Swain also cites the work of several other German scholars, including E. Haselhoff and G. Lindau, "Die Beschadigung der Vegetation durch Rauch" [The Injury of Vegetation by Smoke] (Berlin, Bornträger, 1903); and A. Wieler, "Untersuchungen über die Einwirkung Schwefeliger Säure auf die Pflanzen" [Research into the Effect of Sulfuric Acid on Plants] (Berlin, Bornträger, 1905). Also see Fulton, *Metallurgical Smoke*, 15.

48. Cameron, *Cottrell*, 110–112.

49. Ibid., 117.

50. Joseph Warren Barker, *Research Corporation (1912–1952): Dedicated to Progress in Education and Science* (New York: Newcomen Society in North America, 1952), 8, notes that Cottrell had apparently read of Lodge's work as early as 1890 when he was only thirteen. He also noted that the earliest known suggestion that electrical charges could be used to capture suspended particles was made in 1824 in an article by the German scientist J. Hohlfeld, "Das Niederschlagen des Rauchs durch Electricitat," *Kastner Archiv. Natural.* 2 (1824): 205–206. See Cottrell, "The Electrical Precipitation of Suspended Particles," in *Recent Copper Smelting*, ed. Thomas T. Read (San Francisco: Mining and Scientific Press, 1914), 246.

51. *Encyclopedia of Chemical Technology*, 3rd ed. (New York: John Wiley & Sons, 1977), 673.

52. Barker, *Research Corporation*, 9.

53. Cameron, *Cottrell*, 121–124.

54. Barker, *Research Corporation*, 9–10.

55. Cameron, *Cottrell*, 126–130.

56. Ibid., 128–130.

57. Frederick G. Cottrell, "The Social Responsibility of the Engineer," *Science* 85 (1937): 529–531, quote from 530.

58. See Cottrell's diary and personal notebook, 1 July to 26 August 1907, vol. 1, box 1, Frederick G. Cottrell Papers, Manuscript Division, Library of Congress; *Dictionary of American Biography*; Cottrell, "Electrical Precipitation: Historical Sketch," *Transactions of the American Institute of Electrical Engineers* 34, pt. 1 (1915): 387–396; *National Cyclopedia of American Biography*.

59. Cameron, *Cottrell*, 128.

60. *United States Patent Office, Official Gazette,* Patent Number 895,729, August 11, 1908, 1233–1234.

61. Cameron, *Cottrell,* 128–130.

62. *Encyclopedia of Chemical Technology,* 677–682; Cameron, *Cottrell,* 129–130.

63. Ibid., 144–148.

64. Cottrell also made brief visits to Garfield to supervise the precipitator installation work. Cottrell diary, July 1911, vol. 1, box 1, Frederick G. Cottrell Papers, Manuscript Division, Library of Congress.

65. Linn Bradley, "Practical Applications of Electrical Precipitation and Progress of the Research Corporation," *Transactions of the American Institute of Electrical Engineers* 43 (1915): 425; "The Cottrell Plant at Garfield, Utah," *Engineering and Mining Journal* 98 (1914): 873; W. H. Howard, "Electrical Fume Precipitation at Garfield," *Transactions of the American Institute of Mining Engineers* 44 (1914): 540–560.

66. Cameron, *Cottrell,* 146.

67. Servos, *Physical Chemistry,* 242; Cameron, *Cottrell,* 113, 150, 209.

68. See Edwin T. Layton Jr., *The Revolt of the Engineers: Social Responsibility and the American Engineering Profession* (Baltimore: Johns Hopkins University Press, 1986).

69. Cottrell, "Social Responsibility," 529–531.

70. Frederick Cottrell, "The Research Corporation, an Experiment in Public Administration of Patent Rights," *Journal of Industrial Engineering Chemistry* 4 (1912): 864–867; James S. Coles, "The Cottrell Legacy: Research Corporation, a Foundation for the Advancement of Science," in *Cottrell Centennial Symposium: Air Pollution and Its Impact on Agriculture* (Turlock, CA: Cal State Associates, 1977), viii–xvii.

71. Cottrell, "Social Responsibility," 532–533.

72. Walter A. Schmidt, "The Control of Dust in Portland Cement Manufacture by Cottrell Precipitation Processes," *Eighth Annual Congress of Applied Chemistry* 5 (1912): 117–124; Cameron, *Cottrell,* 146–148.

73. *Encyclopedia of Chemical Technology,* 683. By the 1970s, though, the precipitator business had been established as a separate, publicly owned subsidiary called Research-Cottrell. See Coles, "The Cottrell Legacy," xiv.

74. Patent Number 895,729.

75. "Mineral Losses in Gases and Fumes," in Read, *Recent Copper Smelting,* 227.

76. Cottrell diary, vol. 1, 27 August 1907, Frederick G. Cottrell Papers, Manuscript Division, Library of Congress; Cameron, *Cottrell,* 130–135.

77. Ibid.

78. Frederick Laist to Cottrell, 23 December 1907, and Mathewson to Cottrell, 13 January 1908, box 63, Anaconda Copper Mining Company Records, Montana Historical Society, Helena (hereafter ACMC Records).

79. Cottrell to Mathewson, 17 January 1908, box 63, ACMC Records.

80. F. F. Frick to Frederick Laist, 16 December 1913, box 63, ACMC Records.

81. Arthur E. Wells, "Report of the Anaconda Smelter Smoke Commission," 1 October 1920, 10, National Archives, Record Group 70, box 278.

82. MacMillan, "Struggle to Abate Air Pollution," 157–162; Fulton, *Metallurgical Smoke*, 84–85.

83. MacMillan, "Struggle to Abate Air Pollution," 229.

84. Ibid., 287; J. K. Haywood, *Injury to Vegetation and Animal Life Along Silver Bow and Warm Springs Creek*, Bulletin 113 (Washington: U.S. Department of Agriculture, 1908), 21–36.

85. Ibid., 21–36; MacMillan, "Struggle to Abate Air Pollution," 201–202.

86. A. F. Potter to Joseph A. Holmes, box 84-A, National Archives, Record Group 70.

87. Quoted in MacMillan, *Smoke Wars*, 213.

88. Although the Selby Smoke Commission was created later than the Anaconda commission, the work of the Selby group proceeded much faster, and they published their final report in 1915, well before the Anaconda commission's 1920 report.

89. MacMillan, "Struggle to Abate Air Pollution," 302.

90. Marcosson, *Anaconda*, 256–259; *National Cyclopedia of American Biography*, s.v. "Louis Davidson Ricketts," notes that the Princeton-trained mining engineer was unrivaled in having "contributed in a major way to the success of so many big copper mining companies."

91. *National Cyclopedia of American Biography*, s.v. "Joseph Holmes." Also see vols. 2 and 3 of Mary Rabbitt, *Minerals, Lands, and Geology for the Common Defence and General Welfare: A History of Public Lands, Federal Science and Mapping Policy, and Development of Mineral Resources in the United States*, 3 vols. (Washington, D.C.: U.S. Geological Survey, 1979–1986).

92. *Dictionary of American Biography*, s.v. "Joseph Holmes"; William Graebner, *Coal-Mining Safety in the Progressive Period: The Political Economy of Reform* (Lexington: University Press of Kentucky, 1976), 35–36.

93. Kenneth S. Mernitz, "Governmental Research and the Corporate State: The Rittman Refining Process," *Technology and Culture* 31 (1990): 88. See also *Dictionary of American Biography*, s.v. "Joseph Holmes."

94. "The Director of the Bureau of Mines," *The Engineering and Mining Journal* 89 (1910): 1096.

95. *Engineering and Mining Journal* 89 (1910): 4.

96. MacMillan, "Struggle to Abate Air Pollution," 305. Holmes's Office of Air Pollution was abolished later by H. Van Manning, though the bureau's research efforts on smoke pollution continued. Charles Jackson, *The United States Bureau of Mines* (Washington: Congressional Research Division, 1976), 29; Fred Wilbur Powell, *The Bureau of Mines: Its History and Activities* (New York: D. Appleton, 1922), 21.

97. Wells, *Smelter Smoke*, 19.

98. Cameron, *Cottrell*, 154.

99. Ibid., 154–159.

100. Holmes, *Second Annual Report of the Director of the Bureau of Mines* (Washington, D.C.: Bureau of Mines, 1912), 68; Wells, *Smelter Smoke*, 32; Edgar M. Dunn, "Determination of Gases in Smelter Fumes," and "Notes on the Determination of Dust Losses at the Washoe Reduction Works, Anaconda, Montana," *Transactions of the American Institute of Mining Engineers* 46 (1913): 648–652.

101. Van H. Manning, *Seventh Annual Report of the Director of the Bureau of Mines* (Washington, D.C.: Bureau of Mines, 1917), 79–87.

102. Van H. Manning to Karl Eilers, President of the American Smelting and Refining Company, August 23, 1916, National Archives, Record Group 70, box 223.

103. Frederick Laist to Cottrell, 23 December 1907, and Mathewson to Cottrell, 13 January 1908, box 63, ACMC Records.

104. Swain, "Smoke and Fume," 2385; Wells, *Smelter Smoke*, 25. Like the Garfield smelter, the Washoe also faced difficulties in using bag houses because of the corrosive effects of the sulfuric acid emissions.

105. Cottrell's exacting demands for precise flue dust testing are discussed in J. H. Klepinger to Cottrell, 18 October 1911, box 261, ACMC Records.

106. On the commission's efforts to investigate arsenic recovery practices, see J. O. Elton, "Arsenic Trioxide from Flue Dust," *Transactions of the American Institute of Mining Engineers* 46 (1913): 690. See also Wells, *Smelter Smoke*, 25, 53; and H. Welch and L. H. Duschak, "The Vapor Pressure of Arsenic Trioxide," Technical Paper 81 (Washington, D.C.: Bureau of Mines, 1914).

107. Cottrell diary, 14 August 1913, box 1, Frederick G. Cottrell Papers, Manuscript Division, Library of Congress.

108. Quoted in MacMillan, *Smoke Wars*, 213.

109. Wells, *Smelter Smoke*, 2.

110. Ibid., 55.

111. MacMillan, "Struggle to Abate Air Pollution," 313–314.

112. Wells, *Smelter Smoke*, 50.

113. A. E. Wells to Frederick Cottrell, 20 January 1917, National Archives, Record Group 70, box 302.

114. Wells, *Smelter Smoke*, 30.

115. Anaconda Smelter Commission, "Progress Report," 21 June 1922, National Archives, Record Group 70, box 302.

116. Frederick Laist to Van H. Manning, 20 July 1918, National Archives, Record Group 70, box 84-A.

117. MacMillan, "Struggle to Abate Air Pollution," 332–349. Even in contemporary copper smelting plants designed to capture and utilize pollutants, the levels of arsenic and sulfur compounds that continue to be released have often been dangerously high. For example, ASARCO's Tacoma, Washington, smelter released 282,000 kilograms of arsenic every year until it shut down in 1984, despite the plant's arsenic capture and production technology. See Peter Dorman, "Environmental Protection, Employment, and Profit: The Politics of Public Interest in the

Tacoma/Asarco Arsenic Dispute," *Review of Radical Political Economics* 16 (1984): 151–176.

118. Vannevar Bush, "Frederick Gardner Cottrell," *National Academy of Science Biographical Memoirs* 22, (1952): 1–11, provides a brief overview of Cottrell's career after the Bureau of Mines.

119. *National Cyclopedia of American Biography*, s.v. Cottrell, 294–295.

120. Cameron, *Cottrell*, 133.

121. Fumio Matsumura, *Toxicology of Insecticides* (New York: Plenum Press, 1985), 92–93, notes that arsenic can also be toxic to plant life. See also Hallenbeck, *Pesticides*, 17.

122. Matsumura, *Toxicology of Insecticides*, 92–93; James C. Whorton, *Before Silent Spring: Pesticides and Public Health in Pre-DDT America* (Princeton: Princeton University Press, 1975), 17, observes that arsenical compounds—usually arsenic trioxide—were the most important and widely used of the first inorganic pesticides.

123. Alan S. Newell, "A Brief Historical Overview of Anaconda Copper Mining Company's Principal Mining and Smelting Facilities Along Silver Bow and Warm Spring Creeks, Montana" (Missoula: Montana Historical Research Associates, 1995), 52; Environmental Protection Agency, "Proposed Plan: Rocker Timber Framing and Treating Plant Operable Unit" (Washington, D.C.: EPA, 1995); Joshua Lipton, "Cleaning Up Montana: Superfund Accomplishments" (Helena: Montana Department of Environmental Quality, 1996), 35.

124. Laurie K. Mercier, " 'The Stack Dominated Our Lives': Metal Manufacturing in Four Montana Communities," *Montana: The Magazine of Western History* 38 (1988): 40–57. Particularly deadly was the task of emptying the hoppers beneath the Washoe dust chambers.

125. The Anaconda Company initially stored about three-quarters of the arsenic captured by the precipitators. See Frederick Laist to Cornelius Kelley, 22 September 1914, box 63, ACMC Records. On heavy metal contamination of the Clark Fork, see Environmental Protection Agency, "Superfund Program Fact Sheet: Anaconda Smelter Site, Anaconda, Montana," June 1985. On Mill Creek, see Joshua Lipton, "'Terrestrial Resources Injury Assessment Report" (Helena: State of Montana Natural Resource Damage Program, 1993), section 2, 2–5, and Lipton, "Cleaning Up Montana," 26.

126. Lawrence A. Smith et al., *Remedial Options for Metals Contaminated Sites* (Boca Raton, Fla.: Lewis Publishers, 1995), 17.

FOUR. THE PIT

1. "Summarized History Mining and Milling Developments, Utah Copper Company," C-B 1047, Ringbinder III: 1–14, Bancroft Library, D. C. Jackling Collection (hereafter Bancroft Jackling Collection). Also "Copper Mining Communities and Their People," pamphlet published by Kennecott Utah Copper, c. 1985.

2. See, for example, "Makers of Mine and Markets, Daniel C. Jackling, Who Makes Mole Hills out of Mountains," *Copper Curb and Mining Outlook*, 20 July 1920, Personal Scrapbook, Bancroft Jackling Collection; *Let's Get Acquainted* (Salt Lake City: Kennecott Copper, 1956), 3.

3. Leonard J. Arrington and Gary B. Hansen, *The Richest Hole on Earth: A History of the Bingham Copper Mine* (Logan: Utah State University Press, 1963), 7.

4. T. A. Rickard, "The Utah Copper Enterprise—I: History," *Mining and Scientific Press*, 5 October 1918, 447; Wilbur Smith, "Early Day Bingham and Utah Copper Mining," typed unpublished manuscript, folder 1, box 1, Wilbur Smith Papers, University of Utah Marriott Library, Salt Lake City.

5. This argument is made in Rickard, "Utah Copper Enterprise," 446, and S. F. Emmons, *Economic Geology of the Mercur Mining District, Utah* (Washington, D.C.: U.S. Geological Survey, 1918).

6. Leonard J. Arrington, "Abundance from the Earth: The Beginnings of Commercial Mining in Utah," *Utah Historical Quarterly* 31 (1963): 194; Arrington, *Richest Hole*, 11; Harvey O'Connor, *The Guggenheims: The Making of an American Dynasty* (New York: Covici, Friede, 1937), 275.

7. Quoted in Gary B. Hansen, "Industry of Destiny: Copper in Utah," *Utah Historical Quarterly* 31 (1963): 263.

8. Arrington, "Abundance from the Earth," 200; Lynn Robison Bailey, *Old Reliable: A History of Bingham Canyon, Utah* (Tucson, Ariz.: Westernlore Press, 1988), 16.

9. Hansen, "Industry of Destiny," 265–278.

10. Ibid., 5.

11. Daniel C. Jackling Papers, Genealogy File, box 1, Special Collections, Stanford University Library, Palo Alto, California (hereafter Stanford Jackling Collection).

12. "Cowan, W. B." folder, box 3, Stanford Jackling Collection.

13. Bruce D. Whitehead and Robert E. Rampton, "Bingham Canyon," in *From the Ground Up: The History of Mining in Utah*, ed. Colleen Whitley (Logan: Utah State University Press, 2006), 226–227.

14. See Clarence Nelson Roberts, *History of the University of Missouri School of Mines* ([Rolla?], Mo.: n.p, 1946); Clair Victor Mann, *A Brief History of Missouri School of Mines and Metallurgy* (Rolla, Mo.: N.p., 1939); and *The History of Missouri School of Mines and Metallurgy* (Rolla, Mo.: Phelps County Historical Society, 1941).

15. Larry Gragg, " 'Hell, I Saved the School': George E. Ladd and the Missouri School of Mines and Metallurgy, 1897–1907," University of Missouri–Rolla, available online at http://campus.umr.edu/archives/Pages/Archival Resources/About MSM/ladd.html, accessed 23 June 2007.

16. Bailey, *Old Reliable*, 42–43.

17. Ibid., 42–43; Robert W. Fogel et al., "Secular Change in American and British Stature and Nutrition," *Journal of Interdisciplinary History* 7 (1994): 463; O'Connor, *Guggenheims*, 277.

18. Whitehead, "Bingham Canyon," 227.

19. *National Cyclopedia of American Biography*, s.v. "Joseph Rafael De La Mar." Also see Bailey, *Old Reliable*, 40.

20. T. A. Rickard, *Interviews with Mining Engineers* (San Francisco: Mining and Scientific Press, 1922), 191–199; Arnold Irvine, "The People Who Make Kennecott," *Deseret News*, 27 March 1985.

21. *National Cyclopedia of American Biography*, s.v. "Robert Campbell Gemmell"; "R. C. Gemmell," *Transactions of the American Institute of Mining Engineers* 68 (1923): 1168–1171. On Enos Wall, see Bailey, *Old Reliable*, 40.

22. Rickard, *Interviews*, 204–205.

23. Bailey, *Old Reliable*, 40–41.

24. Rickard, *Interviews*, 204–205.

25. "The Boston Consolidated Copper Mine," *Engineering and Mining Journal* 68 (1899): 614.

26. Parsons, *Porphyry Coppers*, 7.

27. Arrington, *Richest Hole*, 34.

28. Parsons, *The Porphyry Coppers in 1956* (New York: American Institute of Mining Engineers, 1957), 32.

29. Joralemon, *Romantic Copper*, 233–234; Arrington, *Richest Hole*, 35.

30. Helen R. Fairbanks and Charles P. Berkey, *Life and Letters of R.A.F. Penrose, Jr.* (New York: Geological Society of America, 1952), 308.

31. Jeffrey J. Safford, *The Mechanics of Optimism: Mining Companies, Technology, and the Hot Spring Gold Rush, Montana Territory, 1864–1868* (Boulder: University Press of Colorado, 2004).

32. Jackling quoted in *Salt Lake City Herald-Republican*, 6 March 1913.

33. Whitehead, "Bingham Canyon," 226.

34. The Utah historian Leonard Arrington also suggests the idea that Jackling's hard early years may have driven his enduring interest in the Bingham prospect. See Arrington, *Richest Hole*, 36.

35. "Boston Copper Gossip: Metal Market Reported Bare of Copper," *New York Times*, 26 March 1906.

36. Joralemon, *Romantic Copper*, 229.

37. K. E. Porter and D. L. Edelstein, comps., U.S. Geological Survey, "Copper Statistics," available online at http://minerals.usgs.gov/ds/2005/140/copper.pdf, accessed 13 May 2008.

38. "The Copper Collapse," *New York Times*, 22 September 1907.

39. Porter and Edelstein, "Copper Statistics."

40. In his survey of the period, however, Orris C. Herfindahl finds little concrete evidence of a copper pool or other attempts at deliberate price manipulations. He concludes the extraordinary growth in demand for copper was the more likely cause. See *Copper Costs and Prices: 1870–1957* (Baltimore: Johns Hopkins University Press, 1959), 85–88.

41. Joralemon, *Copper*, 229.

42. Porter and Edelstein, "Copper Statistics."

43. The contents of Joralemon's 1934 book were included unchanged in a 1973 book that updated the story with six additional new chapters: Ira Joralemon, *Copper: The Encompassing Story of Mankind's First Metal* (Berkeley: Howell-North Books, 1973), 202–203. Hereafter, the original 1934 book is referred to as *Romantic Copper*, the 1973 volume as *Copper*.

44. Joralemon, *Copper*, 203.

45. Ibid., 223.

46. William T. Parry, "Geology and Utah's Mineral Treasures," in Whitley, *From the Ground Up*, 26–30.

47. Arrington, *Richest Hole*, 83; Joralemon, *Copper*, 231.

48. Joralemon, *Copper*, 229.

49. *National Cyclopedia of American Biography*, s.v. "Spencer Penrose" and "Richard Alexander Fullerton Penrose, Jr."; *Who Was Who in America*, s.v. "Charles Mather MacNeill"; Bailey, *Old Reliable*, 43.

50. Charles K. Hyde, *Copper for America: The United States Copper Industry from Colonial Times to the 1990s* (Tucson: University of Arizona Press, 1998), 140–141.

51. Mike Gorrell, "Miner Who Brought Block Caving to Utah Honored," *Salt Lake City Tribune*, 14 September 2005.

52. Arrington, *Richest Hole*, 90.

53. Ibid., 41–44; Bailey, *Old Reliable*, 47–48.

54. Hyde, *Copper for America*, 141.

55. Porter and Edelstein, "Copper Statistics."

56. Arrington, *Richest Hole*, 41.

57. Ibid., 41–44, 47; Hyde, *Copper for America*, 141; Bailey, *Old Reliable*, 47–48.

58. Joralemon, *Copper*, 235.

59. Ibid., 235.

60. Rossiter W. Raymond to Sarah, 24 August 1872, acc. 1063, Rossiter Worthington Raymond Collection, Colorado Historical Society, Denver.

61. Arrington, *Richest Hole*, 45.

62. Joralemon, *Copper*, viii.

63. Jackling quoted in "Electrical Development Praised: They Marvel at Butte and Superior," *Anaconda Standard*, 13 October 1913.

64. Johns Hays Hammond, foreword to Parsons, *Porphyry Coppers*, x.

65. Arrington, *Richest Hole*, 1.

66. Thomas P. Hughes, *Networks of Power: Electrification in Western Society, 1880–1930* (Baltimore: Johns Hopkins University Press, 1983).

67. Harold Barger and Sam H. Schurr, *The Mining Industries, 1899–1939: A Study of Output, Employment, and Productivity* (New York: Arno Press, 1975), 107.

68. Parsons, *Porphyry Coppers*, 35.

69. For contemporary examples, see the otherwise excellent article by Logan Hovis and Jeremy Mouat, "Miners, Engineers, and the Transformation of Work in

the Western Mining Industry," *Technology and Culture* 37 (1996): 433; and Arrington, *Richest Hole*, 8.

70. Parsons, *Porphyry Coppers*, 1.

71. David A. Hounshell, *From the American System to Mass Production, 1800–1932: The Development of Manufacturing Technology in the United States* (Baltimore: Johns Hopkins University Press, 1984), 1–2.

72. Richard White, *The Organic Machine* (New York: Hill and Wang, 1995).

73. David Igler, *Industrial Cowboys: Miller & Lux and the Transformation of the Far West, 1850–1920* (Berkeley: University of California Press, 2005).

74. See esp. Edmund Russell, "Evolutionary History: Prospectus for a New Field," *Environmental History* 8, no. 2 (2003): 204–228. See also Russell's introductory essay in Philip Scranton and Susan R. Schrepfer, eds., *Industrializing Organisms: Introducing Evolutionary History* (New York: Routledge, 2004).

75. Deborah Fitzgerald, *Every Farm a Factory: The Industrial Ideal in American Agriculture* (New Haven: Yale University Press, 2003).

76. Although he did not deal explicitly with mining, the reference here is to Leo Marx, *The Machine in the Garden: Technology and the Pastoral Ideal in America* (New York: Oxford University Press, 1964).

77. Lindy Biggs, "The Engineered Factory," *Technology and Culture* 36 (Supplement, 1995): S174–S188; Alfred D. Chandler, *The Visible Hand: The Managerial Revolution in American Business* (Cambridge, Mass.: Belknap Press, 1977), 244–283.

78. *Conquering the Earth* (Wilmington, Del.: Hercules Powder, 1923), 24.

79. See also the script for a slide show developed by Canadian Industries Limited and presented to the 1945 Clinic of the DuPont Advertising Department, "Just What C.I.L Adds Up To," acc. no. 395:14, Hagley Museum and Library Manuscripts Department, Wilmington, Delaware.

80. Lewis Mumford, *The Myth of the Machine: The Pentagon of Power* (New York: Harcourt Brace Jovanovich, 1970), 188–211, quotes from 191, 196.

81. *Conquering the Earth*, 24.

82. Stephen George, "The Origins and Discovery of the First Nitrated Organic Explosives" (Ph.D. diss., University of Wisconsin, Madison, 1977), 96.

83. "Gold Mining—From Prospecting Days to the Present," *DuPont Magazine*, 6 June 1928.

84. Otis E. Young, *Black Powder and Hand Steel* (Norman: University of Oklahoma Press, 1976), 37–39.

85. Norman B. Wilkinson, *Lammot du Pont and the American Explosives Industry, 1850–1884* (Charlottesville: University Press of Virginia, 1984), 248; H. C. Chelson, "From Gunpowder to Dynamite," *Engineering and Mining Journal* 137 (1936): 232–233.

86. Ibid., 259; Wilkinson, *Lammot du Pont*, 248–264.

87. *Joplin Daily News*, 21 March 1878, clipping in the P. S. du Pont Office Scrapbook, acc. no. 501:33, Hagley Museum and Library Manuscripts Department, Wilmington, Delaware.

88. Letter from W. C. Spruance to William Coyne, 28 June 1913, acc. no. 518: 1011–27, Hagley Museum and Library Manuscripts Department, Wilmington, Delaware.

89. Barger, *Mining Industries*, 122.

90. William S. Dutton, *DuPont: One Hundred and Forty Years* (New York: Charles Scribner's Sons, 1942), 188–189; Arthur Pine Van Gelder, *History of the Explosives Industry in America* (New York: Columbia University Press, 1927), 611; David A. Hounshell and John Kenly Smith Jr., *Science and Corporate Strategy: Du Pont R&D, 1902–1980* (Cambridge and New York: Cambridge University Press, 1988), 11, 20–26.

91. Report to Hamilton Barksdale from the High Explosives Operating Department on work done by the Eastern Laboratory, 28 August 1914, acc. no. 518: 1011-A/107, Hagley Museum and Library Manuscripts Department, Wilmington, Delaware.

92. Parsons, *Porphyry Coppers*, 375–376.

93. Van Gelder, *Explosives Industry*, 962–973.

94. Parsons, *Porphyry Coppers*, 374–375.

95. Ibid., 372, 377–378; Brian Hayes, *Infrastructure: The Book of Everything for the Industrial Landscape* (New York: Norton, 2005), 21.

96. E. D. Gardner and McHenry Mosier, *Open-Cut Metal Mining*, Bulletin 433 (Washington, D.C.: Bureau of Mines, 1941), 112, 130.

97. Rickard, *Utah Copper Enterprise*, 40.

98. "Map Showing Location of Blasting Powder Plants, 1915" acc. no. 472:127, Hagley Museum and Library Manuscripts Department, Wilmington, Delaware.

99. Rickard, *Utah Copper Enterprise*, 37, 40.

100. Parsons, *Porphyry Coppers*, 355; Barger, *Mining Industries*, 119.

101. *A Brief History of Bucyrus and the Excavating Industry* (South Milwaukee: Bucyrus-Erie, 1947), 13; Larry Lankton, *Cradle to Grave: Life, Work, and Death at the Lake Superior Copper Mines* (New York: Oxford University Press, 1991); idem., "The Machine Under the Garden: Rock Drills Arrive at the Lake Superior Copper Mines, 1868–1883," *Technology and Culture* 24 (1983): 1–37.

102. Charles F. Jackson, *Sampling and Estimation of Ore Deposits* (Washington, D.C.: Bureau of Mines, 1932), 13–17; Parsons, *Porphyry Coppers*, 357–358.

103. Jackson, *Sampling and Estimation of Ore Deposits*, 32, 40; C. E. Nighman and O. E. Kiessling, *Mineral Technology and Output per Man Studies: Rock Drilling* (Philadelphia: Bureau of Mines, 1940), 56–57; James D. Cummings, *Diamond Drill Handbook* (Toronto: J. K. Smith & Sons of Canada, 1956), 30.

104. Parsons, *Porphyry Coppers*, 363.

105. Ibid., 358–360.

106. Hovis, "Western Mining Industry," 441; Parsons, *Porphyry Coppers*, 356.

107. *The Autobiography of John Hays Hammond* (New York: Farrar & Rinehart, 1935), 517.

108. Barger, *Mining Industries*, 115; Jackson, *Sampling and Estimation of Ore Deposits*, 1–12.

109. Clark Spence, *Mining Engineers and the American West: The Lace-Boot Brigade, 1849–1933* (New Haven: Yale University Press, 1970), 244.

110. Hovis, "Western Mining Industry," 441.

111. *John Hays Hammond*, 517.

112. Parsons, *Porphyry Coppers*, 369.

113. Barbara Elleman, *Virginia Lee Burton: A Life in Art* (New York: Houghton Mifflin, 2002).

114. Virginia Lee Burton, *Mike Mulligan and His Steam Shovel* (Boston: Houghton Mifflin, 1939).

115. Parsons, *Porphyry Coppers*, 369.

116. Chandler, *Visible Hand*, 244.

117. Hounshell, *From the American System to Mass Production*, 237.

118. Quoted in Chandler, *Visible Hand*, 265–266.

119. "Obituary: Alfred Chandler, Chronicler of Corporations, Died on May 9th, Aged 88," *Economist*, 17 May 2007.

120. "Harvard Business School Professor Alfred D. Chandler, Jr., Preeminent Business Historian, Dead at 88," Harvard Business School press release, 11 May 2007, available online at http://www.hbs.edu/news/releases/051107_chandler.html, accessed 19 September 2007.

121. See, for example, Paul R. Josephson, *Industrialized Nature: Brute Force Technology and the Transformation of the Natural World* (Washington, D.C.: Island Press/Shearwater Books, 2002).

122. The author is indebted to Andrew Isenberg for suggesting this point.

123. William Cronon, *Nature's Metropolis: Chicago and the Great West* (New York: Norton, 1991).

124. Andrew Isenberg makes this same point in his very insightful recent history of California gold mining and other industries. See *Mining California: An Ecological History* (New York: Hill and Wang, 2005), 15.

125. Chandler, *Visible Hand*, 242.

126. Ibid., 265–266.

127. Ibid., 262–263.

128. Barger, *Mining Industries*, 232. The concept of technological bottlenecks is developed in Hughes, *Networks of Power*.

129. Barger, *Mining Industries*, 232.

130. Ibid., 137. Miners used the terms "open-cut" and "open-pit" interchangeably, although open-cut most commonly applies to relatively flat shallow excavations such as on the Mesabi range.

131. Gardner, *Open-Cut Metal Mining*, 2.

132. *Brief History of Bucyrus*, 1–7.

133. Ibid., 4–6; Harold Francis Williamson and Kenneth H. Myers, *Designed for Digging*, Northwestern University Studies in Business History (Evanston: Northwestern University Press, 1955), 23–25, 27.

134. George B. Anderson, *One Hundred Booming Years: A History of Bucyrus-*

Erie Company (South Milwaukee: Bucyrus-Erie, 1980), and Keith Haddock, *A History of Marion Power Shovel's First 100 Years* (Marion, Ohio: Marion Power Shovel Division, Dresser Industries, 1984), are both uncritical company-sponsored histories, but they offer useful basic historical and technical information.

135. *Why D.C?* (Bucyrus, Ohio: Marion Steam Shovel Company), pamphlet available in mss. collection 1039, box 16, #708, Colorado Historical Society, Denver.

136. *The Right Arm of Progress* (South Milwaukee: Bucyrus-Erie, 1930), 8–9; Williamson and Myers, *Designed for Digging*, 92–99; David McCullough, *The Path Between the Seas: The Creation of the Panama Canal, 1870–1914* (New York: Simon and Schuster, 1977).

137. Gardner, *Open-Cut Metal Mining*, 2.

138. F. E. Cash and M. W. Bernewitz, *Methods, Costs, and Safety in Stripping and Mining Coal, Copper Ore, Bauxite, and Pebble Phosphate*, Bulletin 298 (Washington, D.C.: Bureau of Mines, 1929), 1–5; Grant Holmes, "Early Coal Stripping Full of Heartbreaks," *Coal Age*, 29 May 1924. On the development of Minnesota iron mining, see E. W. Davis, *Pioneering with Taconite* (St. Paul: Minnesota Historical Society, 1964).

139. C. K. Leith, *The Mesabi Iron-Bearing District*, Monograph 43 (Washington, D.C.: U.S. Geological Survey, 1903); C. E. Bailey, "Mining Methods on the Mesabi Range," *Transactions of the American Institute of Mining Engineers* 27 (1897): 529–536; Horace V. Winchell, "The Mesabi Iron-Range," *Transactions of the American Institute of Mining Engineers* 21 (1892–93): 644–686; Margaret Banning, *Mesabi* (New York: Harper & Row, 1969).

140. Arrington, *Richest Hole on Earth*.

141. Parsons, *Porphyry Coppers*, 53.

142. Rickard, *Utah Copper Enterprise*, 47.

143. Rickard, *Interviews with Mining Engineers*, 209.

144. Bailey, *Old Reliable*, 53.

145. Arrington, *Richest Hole*, 52–53; H. C. Goodrich, "Steam Shovel and Ore Shipping at Bingham, Utah," *Mining and Engineering World*, 2 December 1911, 1114–1116.

146. Eugene Delos Gardner and McHenry Mosier, *Open-Cut Metal Mining* (Washington, DC: Government Printing Office, 1941), 126.

147. Barger, *Mining Industries*, 138.

148. Proper grammar would suggest the term should be "fully rotating," but "fully revolving" was the commonly used phrase.

149. Williamson and Myers, *Designed for Digging*, 75.

150. Barger, *Mining Industries*, 139–140; Goodrich, "Shovel Operations at Bingham," 576, 580.

151. Gardner, *Open-Cut Metal Mining*, 11.

152. Rickard, *Utah Copper Enterprise*, 524, 551.

153. H. C. Goodrich, "Shovel Operations at Bingham, Utah Copper Co.," *Transactions of the American Institute of Mining Engineers* 72 (1925): 582; Robert Strong

Lewis, *Elements of Mining* (New York: John Wiley & Sons, 1933), 217–218; Parsons, *Porphyry Coppers*, 381.

154. Williamson and Myers, *Designed for Digging*, 78.

155. Ibid., 78.

156. Niles White, "From Tractor to Tank," *American Heritage of Invention & Technology* (February 1993): 58–64; Walter A. Payne, ed., *Benjamin Holt: The Story of the Caterpillar Tractor* (Stockton, Calif.: University of the Pacific, 1982).

157. E. D. Gardner and C. H. Johnson, *Copper Mining in North America* (Washington, D.C.: Bureau of Mines, 1938), 136.

158. Gardner, *Open-Cut Metal Mining*, 8.

159. G. Townsend Harley, "A Study of Shoveling as Applied to Mining," *Transactions of the American Institute of Mining Engineers* 41 (1920): 147–187.

160. Bucyrus-Erie advertisement in *Civil Engineering*, January 1954.

161. Howard L. Hartman and Jan M. Mutmansky, *Introductory Mining Engineering* (New York: John Wiley & Sons, 2002), 189.

162. Keith Haddock, *Colossal Earthmovers* (Osceola, Wisc.: MBI, 2000), 67.

163. "The World's Largest," *Arch of Illinois News*, February-March, 1985.

164. Chandler, *Visible Hand*, 242.

165. Construction Service Company (New York, *Handbook of Steam Shovel Work*, a report to the Bucyrus Company (New York: Chasmar-Winchell Press, 1911), 350.

166. This rough estimate of value would actually be too small. Assuming a volumetric distribution of 1 to 100, the true amount of gold by weight in the wheelbarrow would be considerably greater than a pound since gold is heavier than sand. The value of a pound of gold here is based on the unusually high market price in late 2007 of about $730 per troy ounce. There are 14.5 troy ounces in a conventional pound.

167. Rickard, *Utah Copper Enterprise*, 51–52.

168. "Concentrating Equipment," Allis-Chalmers 1919 catalog, acc. no. 1039, Morse Brothers Machinery Catalog Collection, Colorado Historical Society, Denver.

169. Young, *Western Mining*, 138–139; Robert H. Richards, *A Textbook of Ore Dressing* (New York: McGraw-Hill, 1925), 220–228.

170. James Colquhoun, *The History of the Clifton-Morenci District* (London: John Murray, 1924), 1–78.

171. *National Cyclopedia of American Biography*, s.v. "Arthur Redman Wilfley."

172. Jay E. Niebur, *Arthur Redman Wilfley* (Denver: Western Business History Research Center of the Colorado Historical Society, 1982), 4–46.

173. Ibid., 4–90.

174. "A Biography of the Mine & Smelter Supply Company," *Mines Magazine* 12, no. 7 (1922): 13–15.

175. Niebur, *Wilfley*, 90–91.

176. "A Biography of the Mine & Smelter Supply Company," *Mines Magazine*, 13–15.

177. Elroy Nelson, "The Mineral Industry: A Foundation of Utah's Economy," *Utah Historical Quarterly* 31 (1963): 184.

178. Though Jackling's steam shovels were not *entirely* nonselective. As discussed previously, by precisely mapping the entire deposit, Jackling understood the broad variations in ore grade in the mine and sought to maintain the optimal grade for his concentrators by mixing the output from different shovels.

179. Rickard, *Utah Copper Enterprise*, 713–714.

180. Arrington, *Richest Hole*, 49–50; Kennecott, "Summarized History Mining and Milling Developments, Utah Copper Company," D. C. Jackling Collection, Ringbinder 3, Bancroft Library, University of California, Berkeley.

181. T. A. Rickard and O. C. Ralston, *Flotation* (San Francisco: Mining and Scientific Press, 1917), 5–11.

182. Ibid., 11–25.

183. Ibid., 33.

184. Ibid., 26–28.

185. Rickard, "The Utah Copper Enterprise—I: History," *Mining and Scientific Press* (5 October 1918): 749.

186. Arrington, *Richest Hole*, 69.

187. Today, however, gold is more commonly concentrated through the controversial process of chemical cyanidation, which has permitted the profitable mining of extremely low-grade ores and is discussed in chapter 5. However, even some cyanidation-based gold concentrators also use flotation as a supplemental process.

188. A. Soderberg, *Mining Methods and Costs at the Utah Copper Co., Bingham Canyon, Utah*, Information Circular 6234 (Washington, D.C.: Bureau of Mines, 1930), 19; Rickard, "Utah Copper Enterprise," 787–789.

189. David Wilburn, Thomas Goonan, and Donald Bleiwas, "Technological Advancement—A Factor in Increasing Resource Use," U.S. Geological Survey Open-File Report 01–197 (2001), 55, available online only at http://pubs.usgs.gov/of/2001/of01–197/, accessed 17 December 2007.

190. Daniel C. Jackling, "The Engineer's Province and Obligation in Organized Society," *Journal of the Western Society of Engineers* 45 (1940): 24–27, quotation on 24.

191. Ibid., 25.

192. *Salt Lake Tribune*, 14 October 1909.

193. Joralemon, *Copper*, 295–297.

194. Porter and Edelstein, "Copper Statistics."

195. Parsons, "An Achievement of Engineers," chapter 1 of *Porphyry Coppers*, 1–18.

196. Millie Robbins, "The Fabulous Life of Daniel Jackling," *San Francisco Chronicle*, 25 February 1959.

197. Jackling, "Engineer's Province," 24.

198. Ibid., 25.

FIVE. THE DEAD ZONES

1. Brian Shovers, "The Perils of Working in the Butte Underground: Industrial Fatalities in the Copper Mines, 1880–1920," *Montana: The Magazine of Western History* 37, no. 2 (Spring 1987): 26.

2. Zena Beth McGlashan, correspondence with author, 14 March 2008.

3. T. A. Rickard and O. C. Ralston, *Flotation* (San Francisco: Mining and Scientific Press, 1917), 11–25.

4. Arthur E. Wells, "Report of the Anaconda Smelter Smoke Commission," 1 October 1920, 35–37, National Archives, Record Group 70, box 278.

5. Ibid., 7–8; Fredric L. Quivik, "Smoke and Tailings: An Environmental History of Copper Smelting Technologies in Montana, 1880–1930" (Ph.D. diss., University of Pennsylvania, 1998), 370; Donald MacMillan, *Smoke Wars: Anaconda Copper, Montana Air Pollution, and the Courts, 1890–1924* (Helena: Montana Historical Society Press, 2000), 241–242.

6. Donald MacMillan, "A History of the Struggle to Abate Air Pollution from Copper Smelters of the Far West, 1885–1933" (Ph.D. diss., University of Montana, 1973)," 332–349.

7. K. E. Porter and D. L. Edelstein, comps., U.S. Geological Survey, "Copper Statistics," available online at http://minerals.usgs.gov/ds/2005/140/copper.pdf, accessed 13 May 2008. Not all of the increased consumption was met by new production from mining, as beginning in 1906 so-called secondary production of scrap and other recycled copper played a growing role. Amounts have been converted from metric tons to short tons: 2,000 pounds per ton. Average price for the 1920s is based on the period from 1923 to 1929.

8. Quivik, "Smoke and Tailings," 436, fn. 106.

9. Donald Dewees, "Sulfur Dioxide Emissions from Smelters: The Historical Inefficiency of legal Institutions," 11 April 1996, unpublished manuscript in author's possession.

10. MacMillan, *Smoke Wars*, 246–247; Quivik, "Smoke and Tailings," 474.

11. MacMillan, *Smoke Wars*, 251.

12. Of course, people still lived and worked in the valley, including the citizens of the smelter town of Anaconda. As Laurie Mercier notes in her history of the town, *Anaconda: Labor, Community, and Culture in Montana's Smelter City* (Urbana and Chicago: University of Illinois Press, 2001), 195–196, by the 1970s tests revealed dangerously elevated levels of arsenic, cadmium, and other potential carcinogens in children's hair and in vegetables from local gardens.

13. Quivik, "Smoke and Tailings," 493.

14. Edwin Layton, *The Revolt of the Engineers: Social Responsibility and the American Engineering Profession* (Baltimore: Johns Hopkins University Press, 1986, 1971).

15. Marcosson, *Anaconda*, 272–280.

16. A similar idea is suggested in Gavin Bridge, "The Social Regulation of

Resource Access and Environmental Impact: Production, Nature, and Contradiction in the US Copper Industry," *Geoforum* 31 (2000): 241.

17. "Butte, One of the World's Greatest Mining Areas," *Anaconda Copper Trailsman* 1, no. 1 (August 15, 1956).

18. Sherwood Ross, "How the United States Reversed Its Policy on Bombing Civilians," *Humanist* (July/August 2005): 19–20.

19. Quoted in ibid., 19.

20. Ibid., 17–19. See also the articles collected in Paul Addison and Jeremy A. Crang, eds., *Firestorm: The Bombing of Dresden* (New York: Ivan R. Dee, 2006), especially the chapter by Sönke Neitzel, "The City Under Attack."

21. Stephen L. McFarland, *America's Pursuit of Precision Bombing, 1910–1945* (Washington, D.C.: Smithsonian Institution Press, 1995).

22. Quoted in Ross, "How the United States Reversed Its Policy on Bombing Civilians," 22. On the complex and controversial issues regarding the dropping of the atomic bombs, see J. Samuel Walker, *Prompt and Utter Destruction: President Truman and the Use of the Atomic Bombs Against Japan* (Chapel Hill: University of North Carolina Press, 1997).

23. "Archbishop's Appeal," *Times* (London), 28 December 1937, 9.

24. Edmund Russell, *War and Nature: Fighting Humans and Insects with Chemicals from World War I to Silent Spring* (Cambridge: Cambridge University Press, 2001), 170–171.

25. Ibid., 117. See also John W. Dower, *War Without Mercy: Race and Power in the Pacific War* (New York: Pantheon Books, 1986).

26. Russell, *War and Nature*, 170–71.

27. Ibid., 106.

28. Jörg Friedrich, *The Fire: The Bombing of Germany, 1940–1945* (New York: Columbia University Press, 2006), 83.

29. On technological determinism, see Langdon Winner, *The Whale and the Reactor: A Search for Limits in an Age of High Technology* (Chicago: University of Chicago Press, 1986); idem, *Autonomous Technology: Technics-out-of-Control as a Theme in Political Thought* (Cambridge: MIT Press, 1977); Merritt Roe Smith and Leo Marx, *Does Technology Drive History? The Dilemma of Technological Determinism* (Cambridge: MIT Press, 1994).

30. For a few key social constructivist challenges to technological determinism, see Donald A. MacKenzie, *Inventing Accuracy: An Historical Sociology of Nuclear Missile Guidance*, Inside Technology (Cambridge: MIT Press, 1990); Donald A. MacKenzie and Judy Wajcman, *The Social Shaping of Technology: How the Refrigerator Got Its Hum* (Milton Keynes and Philadelphia: Open University Press, 1985); Wiebe E. Bijker, Thomas Parke Hughes, and T. J. Pinch, *The Social Construction of Technological Systems: New Directions in the Sociology and History of Technology* (Cambridge: MIT Press, 1987).

31. Ronald Schaffer, *Wings of Judgment: American Bombing in World War II* (New York: Oxford University Press, 1985), 162.

32. "Archbishop's Appeal."

33. Anaconda Company advertisement, "See America the Bountiful," *Saturday Review*, 27 July 1957, inside front cover.

34. Ibid.

35. The Stanford historian David Potter popularized this phrase only a year later in his 1958 book, *People of Plenty: Economic Abundance and the American Character* (Chicago: University of Chicago Press, 1958).

36. Gary Cross, *An All-Consuming Century: Why Commercialism Won in Modern America* (New York: Columbia University Press, 2000).

37. This is a contemporary term used in Lizabeth Cohen, *A Consumer's Republic: The Politics of Mass Consumption in Postwar America* (New York: Knopf, 2003).

38. R. L. Agassiz, "Review of the Copper and Brass Research Association," *Mining Congress Journal* 17 (1931): 467.

39. "Will the U.S. Run Short of Metals?" Anaconda pamphlet, n.d., c. 1952, Montana Historical Society, Vertical Files, "Butte, Montana."

40. Cohen, *Consumer's Republic*; Alan Brinkley, *The End of Reform* (New York: Knopf, 1995).

41. Cohen, *Consumer's Republic*, 125.

42. Ibid., 127.

43. Anaconda Copper & Brass advertisement, "Copper Serves" (1948), publication unknown, author's collection. The author has purchased many of these ads from antique dealers and other sources, and often the only provenance information is the date of publication.

44. Anaconda Copper & Brass advertisements, "From Radiator to stop light, nothing serves like Copper" (1948), and "In the world of transportation, nothing serves like Copper" (1945), publications unknown, author's collection.

45. The American Brass Company advertisement, "Why Copper Is the Key Metal in So Many Postwar Plans" (1944), publication unknown, author's collection.

46. Anaconda Copper & Brass advertisement, "Want to Be Surprised?" (1945), publication unknown, author's collection.

47. Cohen, *Consumer's Republic*, 195–197.

48. Ibid., 112–113, 121–123.

49. Porter and Edelstein, "Copper Statistics."

50. Charles K. Hyde, *Copper for America: The United States Copper Industry from Colonial Times to the 1990s* (Tucson: University of Arizona Press, 1998), 177.

51. Kennecott Copper Corporation, "in electronics . . . no substitute can do what copper does!" advertisement, *Saturday Evening Post*, July 27, 1957.

52. Anaconda advertisement, "Copper . . . helps freedom ring" (1950), publication unknown, author's collection.

53. "How many workshops to make a howitzer," Anaconda Company advertisement, *Saturday Evening Post*, 20 October 1951; "How Many Bathrooms in a Jet

Plane's Engine?" Anaconda Company advertisement, *Saturday Evening Post*, 29 September 1951.

54. Marcosson, *Anaconda*, 340–341.

55. T. A. Rickard, "The Utah Copper Enterprise—I: History," *Mining and Scientific Press*, 5 October 1918, 524.

56. Ibid., 516.

57. Otto P. Chendron, *Thoughts and Meditations upon Beholding the Utah Copper Mine* (Salt Lake City: Otto P. Chendron, 1943), unpaginated.

58. F. Keith Stepan, "Visitors Center for Bingham Canyon, Utah" (architecture thesis, University of Utah, 1965), 19, 37–38.

59. Anaconda, "See America the Bountiful."

60. *Anaconda Copper Trailsman* 1, no. 1 (August 15, 1956): front page.

61. *Anaconda Copper Trailsman* 7, no. 1 (June 1965): front page.

62. "We All Depend on Mining," *Anaconda Copper Trailsman* 14, no. 1 (1974): back page.

63. *Preliminary Water Balance for the Berkeley Pit and Related Underground Mine Workings, Final Report* (Denver: Camp Dresser & McKee, 1988), 1.

64. *Draft Baseline Risk Assessment, Mine Flooding Operable Unit, Silver Bow Creek/Butte Area NPL Site, Butte, Montana* (Helena: Montana Environmental Protection Agency, 1993), 7–10.

65. Edwin Dobb, "Pennies from Hell: In Montana the Bill for America's Copper Comes Due," *Harper's*, October 1996, 49. See also Joralemon, *Romantic Copper*, 40, 48, 50.

66. Dobb, "Pennies from Hell," 49.

67. "Cleanup at the Berkeley Pit: Plans and Progress" (Anaconda: Atlantic Richfield Company, 1993); *Resource Recovery Project Demonstration #2: Pulsed Plasma Treatment of Berkeley Pit Water* (Butte: U.S. Department of Energy, 1995).

68. Peter Samuel, "Treasure House or Pollution Pit?" *Forbes* 154 (12 September 1994): 54; Keith Schneider, "New Approach to Old Peril: Abandoned Mines in West," *New York Times*, 27 April 1993.

69. *Kennecott South Fact Sheet* (Denver: Environmental Protection Agency, 1997), 1. Jim Woolf, "Water Cleanup Near Kennecott May Cost $2.2 Billion," *Salt Lake Tribune*, 22 January 1993. The EPA notes that early gold, silver, lead, zinc, and copper miners contributed to this waste production before the Bingham Canyon pit had begun.

70. *Kennecott North Fact Sheet* (Denver: Environmental Protection Agency, 1997), 3.

71. *Kennecott South*, 2.

72. Samuel, "Treasure House," 58.

73. Chris Jorgenson, "Kennecott Will Shift Tailings Expansion Beyond S.L. Boundary," *Salt Lake Tribune*, 16 March 1994.

74. "Top 20 Polluters," *High Country News*, 16 September 1996.

75. "An Off-the-Books Polluter," *High Country News*, 16 September 1996. How-

ever, since the TRI makes no distinction between types or toxicity of waste material, the waste produced by Utah Copper is not necessarily directly comparable in environmental impact with that of the Magnesium Corporation.

76. Harold Barger and Sam H. Schurr, *The Mining Industries, 1899–1939: A Study of Output, Employment, and Productivity* (New York: Arno Press, 1975), 90.

77. Bureau of Mines, *1952 Materials Survey*, S-4, VI-39, 47; Logan Hovis and Jeremy Mouat, "Miners, Engineers, and the Transformation of Work in the Western Mining Industry," *Technology and Culture* 37 (1996): 443–445.

78. Bureau of Mines, *1952 Materials Survey*, VI-38–50.

79. Alfred D. Chandler, *The Visible Hand: The Managerial Revolution in American Business* (Cambridge, Mass.: Belknap Press, 1977), 347.

80. William Cronon, *Nature's Metropolis: Chicago and the Great West* (New York: Norton, 1991).

81. Barger, *Mining Industries*, 238.

82. Ibid., n. 21, 394.

83. Kennecott Copper Corporation, *The Utah Copper Story* (Salt Lake City: Kennecott, 1960).

84. See, for example, "New Formula Releases Gold from Sea, Scientists Told: Man Promised 'Open Sesame' to Mineral Riches Within Ten Years—Supply Unlimited," *Washington [D.C.] Star*, 26 March 1934; John Seabrook, "Invisible Gold," *New Yorker*, 24 April 1989, 45; "Gold in Sea Water—Not Enough to Get Rich," *New Scientist*, 7 July 1989, 27.

85. John Robert McNeill, *Something New Under the Sun: An Environmental History of the Twentieth-Century World* (New York: Norton, 2000), 32.

86. E. D. Gardner and McHenry Mosier, *Open-Cut Metal Mining* (Washington, D.C.: Bureau of Mines, 1941), 2.

87. Duane A. Smith, "On the Move Again: Hard Rock Mining in the West," *Montana* 39, no. 1 (1989): 59–63; Seabrook, "Invisible Gold," 69–73; Richard Manning, "Going for the Gold," *Audubon*, January–February 1994; "The New Gold Rush," *U.S. News & World Report*, 28 October 1991, 44–47.

88. Seabrook, "Invisible Gold," 73.

89. Ibid., 61–62.

90. James Brooke, "For U.S. Miners, the Rush is on to Latin America," *New York Times*, 17 April 1994; Joel Millman, "The Treasure of the Sierra Madre," *Forbes* 20 (December 1993): 198–200; Elisabeth Malkin, "Driving to the Gold Rush," *New York Times*, 20 February 2008; William Yardley, "In High Prices, Moribund Mines Find a Silver Bullet," *New York Times*, 3 April 2008.

91. Gardner, *Open-Cut Metal Mining*, 2.

92. "Most Requested Statistics—U.S. Coal Industry," National Mining Association, available online at http://www.nma.org/pdf/c_most_requested.pdf, accessed 18 February 2008.

93. Penny Loeb, "Shear Madness," *U.S. News and World Report*, August 11, 1997, 26–32.

94. Paul J. Nyden, "From Pick and Shovel to Mountaintop Removal: Environmental Justice in the Appalachian Coalfields," *Environmental Law* 34 (2004): 21–86.

95. Ibid., 22–23.

96. Loeb, "Sheer Madness," 26–27.

97. Joby Warrick, "Appalachia Is Paying Price for White House Rule Change," *Washington Post*, 17 August 2004, A1; idem, "Flattening Appalachia's Mountains: Despite Opposition, a White House Rule Change Gives the Coal Industry Free Rein," *Washington Post National Weekly Edition*, 6–12 September 2004, 11.

98. Jim Motavalli, "Once There Was a Mountain: Ravaging West Virginia for 'Clean Coal,'" *E*, November–December 2007, 38. On the environmental battles, see Michael Shnayerson, *Coal River* (New York: Farrar, Straus and Giroux, 2008). See also Shirley Stewart Burns, *Bringing Down the Mountains: The Impact of Mountaintop Removal Surface Coal Mining on Southern West Virginia Communities, 1970–2004* (Morgantown: West Virginia University Press, 2007).

99. Richard Rajala, *Clearcutting the Pacific Rain Forest: Production, Science, and Regulation* (Vancouver: UBC Press, 1998), 7–20.

100. Ibid., 14–30. In other regions, tractor-like skidders predominated after World War II, but the effect was similar. See Peter MacDonald and Michael Clow, "What a Difference a Skidder Makes: The Role of Technology in the Origins of the Industrialization of Tree Harvesting Systems," *History and Technology* 19 (2003): 127–149.

101. McNeill, *Something New Under the Sun*, 307–308; Rajala, *Clearcutting the Pacific Rain Forest*, 32.

102. Ibid., 48.

103. Ibid., 100, 218.

104. Dale A. Burk, *The Clearcut Crisis: Controversy in the Bitterroot* (Great Falls, Mont.: Jursnick Printing, 1970), 22.

105. Rajala, *Clearcutting the Pacific Rain Forest*, 218–219.

106. Paul W. Hirt, *A Conspiracy of Optimism: Management of the National Forests Since World War Two*, Our Sustainable Future (Lincoln: University of Nebraska Press, 1994), xxii.

107. Ibid., xxii–xxiii.

108. Callum Roberts, *The Unnatural History of the Sea* (Washington, D.C.: Island Press/Shearwater Books, 2007), 168. See also Charles Clover, *The End of the Line: How Overfishing Is Changing the World and What We Eat* (Berkeley: University of California Press, 2008).

109. Roberts, *The Unnatural History of the Sea*, 130–135.

110. Ibid., 146–147.

111. Ibid., 153–156.

112. Ibid., 189, 201–203. On the Soviet development of factory trawlers, see also Paul R. Josephson, *Industrialized Nature: Brute Force Technology and the Transformation of the Natural World* (Washington, D.C.: Island Press/Shearwater Books, 2002).

113. Quoted in Roberts, *Unnatural History of the Sea*, 203.

114. McNeill, *Something New Under the Sun*, 243–50.

115. Clover, *End of the Line*.

116. Andrew Isenberg makes a similar point in his penetrating book *Mining California: An Ecological History* (New York: Hill and Wang, 2005), suggesting that the timber, farming, and ranching interests in California were rooted in the state's earlier experience of environmentally destructive hydraulic mining.

117. "Mr. Jackling Receives the John Fritz Medal," *Mining and Metallurgy* 14 (1933): 261–265.

EPILOGUE. FROM NEW DELHI TO THE NEW WEST

1. Somini Sengupta, "Thirsting for Energy in India's Boomtowns and Beyond," *New York Times*, 2 March 2008; idem, "Inside Gate, India's Good Life; Outside, the Servant's Slums," *New York Times*, 9 June 2008. On Ericsson's use of copper for broadband, see the article "Copper Fights Back" on the company's Web site, http://www.ericsson.com/ericsson/news/newsletter/2007/january/060125_copper_fights.shtml.

2. Sengupta, "Thirsting for Energy."

3. W. David Menzie, John H. De Young Jr., and Walter G. Steblez, "Some Implications of Changing Patterns of Mineral Consumption," U.S. Geological Survey, Open-File Report 03–382, available online only at http://pubs.usgs.gov/of/2003/ofo3–382/ofo3–382.html, accessed 23 March 2008.

4. W. David Menzie, United States Geological Survey, "Testimony Before the Committee on Resources, Subcommittee on Energy and Mineral Resources, U.S. House of Representatives," 18 May 2006, available online at http://www.usgs.gov/congressional/hearings/testimony_menzie_18may06.asp, accessed 23 March 2008.

5. Environmental Protection Agency, "Mine Waste Technology Program, 2001 Annual Report," available online at http://www.epa.gov/nrmrl/std/mtb/mwt/annual/annual2001/mwtp2001.pdf, accessed, 12 December 2007.

6. Menzie, "Testimony Before the Committee on Resources, Subcommittee on Energy and Mineral Resources."

7. Donella H. Meadows et al., *The Limits to Growth* (New York: Signet, 1972).

8. Menzie, "Some Implications for Changing Patterns of Mineral Consumption." Also see the very useful history of the debate in John E. Tilton, *On Borrowed Time? Assessing the Threat of Mineral Depletion* (Washington, D.C.: Resources for the Future, 2003).

9. Paul R. Ehrlich, *The Population Bomb* (New York: Ballantine Books, 1968).

10. Thomas Homer-Dixon, "The End of Ingenuity," *New York Times*, 29 November 2006.

11. K. E. Porter and D. L. Edelstein, comps., United States Geological Survey, "Copper Statistics," available online at http://minerals.usgs.gov/ds/2005/140/copper.pdf, accessed 13 May 2008.

12. Homer-Dixon, "End of Ingenuity." See also Thomas F. Homer-Dixon, *The Upside of Down: Catastrophe, Creativity, and the Renewal of Civilization* (Washington, D.C.: Island Press, 2006).

13. R. B. Gordon, M. Bertram, and T. E. Graedel, "Metal Stocks and Sustainability," *Proceedings of the National Academy of Sciences* 103 (2006): 1209–1214.

14. Quoted in David Cohen, "Earth's Natural Wealth: An Audit," *New Scientist*, 23 May 2007, 34–41.

15. This point is convincingly argued in Tilton, *On Borrowed Time.*

16. John Cramer, "Into the Breach—Clark Fork, Blackfoot Rivers Punch through Milltown Dam," *Missoulian*, 29 March 2008, available online at www .missoulian.com/articles/2008/03/29/news/top/news01.txt, accessed 17 June 2008.

17. Ibid.

18. Environmental Protection Agency, *Revised Community Relations Plan: Milltown Reservoir Sediments Superfund Site, Milltown, Montana* (Washington, D.C.: EPA, 1992), section 2, 1–2; William W. Woessner, *Arsenic Sources and Waste Supply Remedial Action Study, Milltown, Montana* (Helena: Montana Department of Health and Environmental Sciences, 1984), i, 1–2.

19. Perry Backus, "Shipping Sediment: First Loads of Contaminated Material Leave Milltown," *Missoulian*, 3 October 2007; idem, "Milltown Sediment Starts to Fill Tailings Ponds at Opportunity," *Missoulian.Com Magazine*, November 2007, available online at http://www.missoula.com/news/, accessed 6 December 2007.

20. Jim Robbins, "In a Town Called Opportunity, Distress over a Dump," *New York Times*, 24 August 2005; Backus, "Milltown Sediment Starts to Fill Tailings Ponds at Opportunity."

21. John Cramer, "First Fish in 100 Years Swims Past Dam Site," *Missoulian*, 10 August 2008, available online at www.missoulian.com/articles/2008/04/09/bnews/br35.txt, accessed 17 June 2008 .

22. Pat Williams, "When a Dam Is Demolished, the Old Ways Return," *High Country News* (28 April 2008).

23. Amit Kapur and T. E. Graedel, "Copper Mines Above and Below the Ground," *Environmental Science & Technology* (15 May 2006): 3135–3141.

24. Indeed, Gordon, Bertram, and Graedel ("Metal Stocks and Sustainability," 1212) note that current estimates of global lithospheric copper deposits are based primarily on low-grade porphyry deposits like those at Bingham.

25. A notable exception was a brilliant and moving photo essay by the geographer and author Caitlin DeSilvey: "Watershed Moment: The Death of a Montana Dam," *Slate*, 17 April 2008, available online only at http://www.slate.com/id/2188979, accessed 6 May 2008.

26. James Brooke, "West Celebrates Mining's Past, but Not Its Future," *New York Times*, 4 October 1998.

27. William Langewiesche, "The Profits of Doom," *Atlantic Monthly*, April 2001, 56–62.

28. William Yardley, "In High Prices, Moribund Mines Find a Silver Bullet," *New York Times*, 3 April 2008.

29. Porter and Edelstein, "Copper Statistics."

30. Stephen E. Kesler, "Mineral Supply and Demand into the 21st Century," in *Proceedings, Workshop on Deposit Modeling, Mineral Resource Assessment, and Sustainable Development*, ed. Joseph A. Briskey and Klaus J, Schulz, U.S. Geological Survey Circular 1294 (Washington, D.C: USGS, 2007), 56.

31. Horace Dunbar, *Marcy's Mill* (San Diego: Watson-Jones, 1944), 4.

Index

Index

ARCO. *See* Atlantic Richfield Company
Arizona, copper mining in, 93, 206
arsenic: as beauty aid, 68–69; electrical precipitator and, 82–102 passim; insecticides, 103, 245n122; at Milltown, MT, 225–226; smelting byproduct, 69, 82, 97; timber preservative, 103–104
Atlantic Richfield Company (ARCO), 202–204, 225–228
atomic bombs, 184

Bacchus, UT, 143
bag houses, 76, 84, 97
Balaklala smelter, 86–87
Berkeley Pit, 15–18, *16*, 173–174, 188–206, *189*; abundance from, 179–180, 199–202; advertising of, 188–202; ARCO, 202–204; Bingham Pit and, 174; as dead zone, 180–181, 202–203; development, 178–182, *179*; expansion, 202; flooding, *16*, 16–17, 202–204; as high modernist artifact, 181–182; pollution, 202–204; snow geese deaths, 16–17; Superfund site, 202–204; technology of, 200–201; tourism, 188–202, 205–206
Bielenberg, Nick, 24–25, 63, 70, 224–225
Bingham Pit, *12*, 108–171; Arthur mill, 168–169, 174; Bingham Canyon, 111–112; blasting, 142–143; concentrators, 160–168; Copperton mill, 125; dynamite used at, 139–143; early criticisms of, 117–120; drill sampling, 144–146; electrification, 30; geology, 115–117, 123–124, 144–146; as "the Hill," 108–111; and Jackling, Daniel Cowen, 115–171; Magna mill, 128, 161–168, *167*, 205; map of, *127*; pollution, 204–205; production, 106–107, 128, 179–180; scale, 11–13, 110–111; steam shovels, 128, 146–159; tailings, 165; tourism, *12*, 12–13, 199–200. *See also* Jackling, Daniel Cowen; mass destruction; Utah Copper Company

block caving, 125
Braun, Steven, 43
Brinkley, Alan, 192
brute force technology, 232n12
Bucyrus Company, 152–159 passim
Burton, Virginia Lee, 146–147
Bush, Vannevar, 185–186
Butte, MT, 35–36; Berkeley Pit, *16*, 16–17, 178–182, 188–202; Butte Miners Union, 48–50; health problems, 68–69; Mount Moriah Cemetery, 172–174, 225; murder of Frank Little, 50; smelting, 68–69; smoke pollution, 68–69; Superfund designation, 17, 202–204; Western Federation of Miners, 49. *See also* Anaconda Copper Mining Company (underground mining operations); Berkeley Pit
Butte & Superior Mining Company, 166–167
Butte Miners Union, 48–50

California: Gold Rush of 1849, 38; hydraulic mining, 38–39. *See also* Comstock Lode, NV
Carlin Trend, 209–210
Carson, Rachel, 61
Chandler, Alfred, 129, 147–150, 207
Chendron, Otto P., 199–200
Clark, William Andrews, 41, 68–69, 118, 224
Club of Rome, 222
coal mining, 210–211
Cohen, Lizabeth, 192–195
Colorado School of Mines, 56–60
Colquhoun, James, 122, 162
Comstock Lode, NV, 39–41; accidents, 41; technological innovations, 40–41
concentrators, 160–168; crushing, 162–163, *163*; flotation, 166–169, *167*; gravity concentration, 160–166; slime, 161–162, 168; tailings, 165; vanners, 162; Wilfley table, 162–164
Conquering the Earth (DuPont), 138

Index

Index

About the Author

Timothy J. LeCain is an assistant professor of history at Montana State University. He has published widely on the history of mining and the environment in the American West and frequently serves as an expert witness in mining Superfund cases. A native of Montana, he now lives in Bozeman with his wife and their two children.